CAMBRIDGE COUNTY GEOGRAPHIES

General Editor: F. H. H. GUILLEMARD, M.A., M.D.

T0352319

WILTSHIRE

Cambridge County Geographies

WILTSHIRE

by

A. G. BRADLEY

With Maps, Diagrams and Illustrations

Cambridge:

at the University Press

1909

CAMBRIDGE UNIVERSITY PRESS
Cambridge, New York, Melbourne, Madrid, Cape Town,
Singapore, São Paulo, Delhi, Mexico City

Cambridge University Press
The Edinburgh Building, Cambridge CB2 8RU, UK

Published in the United States of America by Cambridge University Press, New York

www.cambridge.org
Information on this title: www.cambridge.org/9781107621626

First published 1909
First paperback edition 2013

A catalogue record for this publication is available from the British Library

ISBN 978-1-107-62162-6 Paperback

PREFACE

THE author takes this opportunity of acknowledging the invaluable assistance rendered him, particularly in the archaeological, natural history, and geological sections, by the Rev. E. H. Goddard of Clyffe Pypard, editor of the *Wiltshire Archaeological and Natural History Magazine.* He has also to thank Mr Robert W. Merriman of Marlborough, Clerk to the Wilts County Council, as well as some other members of the staff at Trowbridge, for information kindly given regarding details connected with that department.

A. G. BRADLEY.

1 *July*, 1909.

CONTENTS

ILLUSTRATIONS

MAPS

The illustrations reproduced on pp. 3, 5, 9, 11, 13, 17, 19, 23, 27, 30, 36, 38, 39, 42, 47, 53, 55, 59, 65, 71, 74, 79, 81, 85, 89, 96, 97, 98, 102, 103, 108, 109, 114, 143, and 144 are from photographs by Messrs F. Frith & Co., Ltd., of Reigate; those on pp. 21, 87, 90, and 111 are from photographs by E. H. Roberts, of Marlborough; and the illustrations on pp. 15, 44, 50, 57, 130, 131, 134, 135, 148, and 152 are reproduced from photographs taken by Messrs Protheroe & Simons, of Swindon. Mr G. H. Dunmore, of Downton, supplied the views on pp. 25, 60, 115, 117, 139, and 149; Mr J. Cook, of Ilford, those on pp. 7, 45, 62, and 101; and Messrs Tomkins & Barrett, of Swindon, those on pp. 69, 70, and 142. The author is indebted to Mr G. Hemins, of Swindon, for the views on pp. 67 and 132; to Messrs H. W. Taunt & Co., of Oxford, for those on pp. 119 and 137; and to Dr Guillemard, the General Editor of the Series, for the picture on p. 92.

1. County and Shire. Meaning of the Words.

If we take a map of England and contrast it with a map of the United States, perhaps one of the first things we shall notice is the dissimilarity of the arbitrary divisions of land of which the countries are composed. In America the rigidly straight boundaries and rectangular shape of the majority of the States strike the eye at once; in England our wonder is rather how the boundaries have come to be so tortuous and complicated—to such a degree, indeed, that until recently many counties had outlying islands, as it were, within their neighbours' territory. We may guess at once that the conditions under which the divisions arose cannot have been the same, and that while in America these formal square blocks of land, like vast allotment gardens, were probably the creation of a central authority, and portioned off much about the same time; the divisions we find in England own no such simple origin. Our guess would not have been wrong, for such, in fact, is more or less the case. The formation of the English counties in many instances was (and is—for they have altered up to to-day) an affair of slow growth.

King Alfred is credited with having made them, but inaccurately, for some existed before his time, others not till long after his death, and their origin was—as their names tell us—of very diverse nature.

Let us turn once more to our map of England. Collectively, we call all our divisions counties, but not every one of them is accurately thus described. Cornwall, for example, is not. Some have names complete in themselves, such as Kent and Sussex, and we find these to be old English kingdoms with but little alteration either in their boundaries or their names. To others the terminal *shire* is appended, which tells us that they were *shorn* from a larger domain—*shares* of Mercia or Northumbria or some other of the great English kingdoms. The term county is of Norman introduction,— the domain of a *Comte* or Count.

Wiltshire, in Saxon times the clan country of the Wilsoetas, became one of the counties of the Saxon kingdom of Wessex, and had for its capital the town of Wilton, which though now of small size and importance was the capital of Wessex when Alfred the Great was its king. To go further back, Wilton was named from the river Wily on whose banks it stands, Wily-town and so Wil-ton. The boundaries of Wiltshire are almost the same to-day as they were in the eighth century and in the later time of King Alfred, when together with Berkshire it formed the most easterly portion of Wessex, which for some time was the dominant Saxon kingdom and to a certain extent exercised sway over the others.

Salisbury Cathedral from the N.E.

2. General Characteristics. Position and Natural Conditions.

Wiltshire is almost entirely an agricultural and pastoral county. It is not so prominent among English counties as it was before the eighteenth century. This is due to many causes, but chiefly to that immense growth of manufacturing industries which has so greatly changed the character of England, and to a certain extent even that of its people, in the last hundred and fifty years. The greatest development in this respect has been in the Midlands and in the North, where coal and iron are most plentiful, and in certain maritime counties near to these that have also the advantage of sea ports, whence they can not only ship their manufactures abroad but import the material, such as wool and cotton, from which to make them. This immense increase in wealth and population has also caused the transformation of vast tracts of country, once moor, marshland, or wild forest, into fertile farming countries. Many towns too in the South, more especially London, as well as many districts, have for various reasons, easy enough to give if it were necessary, shared to the full in this great development. Every county has of course progressed more or less with the times. But by comparison some have changed very little and remained agricultural counties, which in a country like England whose chief national wealth is now in trade and manufactures means taking a secondary rank. Wiltshire is one of such counties. This falling away in

Marlborough, from Granham Hill

importance is no reproach to Wiltshire or its people. It is due merely to its situation as an inland county without coal or iron to speak of, and to its being too far from London for the huge population of that city to spread their residences and wealth about it as they have over great parts of Essex, Herts, Kent, Surrey, Sussex and even Berkshire. Some people think it an advantage to a county not to be smirched with the smoke of factories, nor cut up into villa residences, nor sprinkled with towns of monotonously-built terraces and streets where people live packed together and repair in a body by train to a big city at a distance every day to earn a livelihood.

No county represents Old England without its many disadvantages more thoroughly than Wiltshire does to-day. Save for a few local industries and one great railroad depot it is agricultural and pastoral from end to end. It is not good from any point of view that England should be one vast workshop unable to feed more than a dwindling fraction of its people. Agriculture is still the cleanest and finest of all pursuits and breeds the best men. That Wiltshire from its suitability for every form of rural industry and country pursuits once held high place, and is now of comparatively small importance in wealth and population because nature has ordained that it shall hold to the oldest and finest of industries, is not a matter for regret. It may even be something to be thankful for. At any rate we only mention this changed position of Wiltshire here as an interesting fact.

Another thing, perhaps, has altered its comparative position still more. Between the fourteenth and the

eighteenth centuries and before the rise of the great
Northern manufacturing towns, portions of Wiltshire
were the principal seat of the English cloth trade,
originally introduced there, as it was into Norfolk and
other districts, by the Flemings. This was a small and
homely business compared to the huge factories which

Bremhill Church

now chiefly monopolise it and which practically killed
(though a little still survives) the Wiltshire trade early
in the nineteenth century. Still it was important having
regard to the small population of England at that
period. It kept many towns that we now consider
small but which then ranked high as busy and prosperous,

and it flourished in Wiltshire because of the great numbers of sheep which were fed on its downs and on the Cotswold hills upon its northern boundary. There were no convenient means of inland transport in those days, neither railroads nor canals, while the high roads were mere tracks, almost impassable in winter and rough and rutty in summer, so that a manufacturer had to plant his business in the district where his material was grown. Wool is now imported from every part of the world and is manufactured on a vast scale by elaborate machinery driven by coal. The fleeces of a single English county like Wiltshire (no longer preeminent even in wool owing to the immense reclamation of once almost worthless lands in other districts) are but an insignificant item in clothing the millions of people at home and abroad that the British manufacturer now clothes. Yet more, the whole wool crop of Britain is but a small portion of that demanded by the great factories of the North. In former days when Wiltshire was a leading wool-producing county British wool not merely clothed the people of England, but so much was sent abroad to foreign manufactures as to constitute a leading source of the wealth of our English people. From this it will be readily understood why the position of Wiltshire among the forty-four English counties has so vastly altered. There are now several towns in Lancashire containing more people than the whole of Wiltshire. In the middle ages and for long afterwards Lancashire was a thinly populated, backward, almost barbarous country compared to Wiltshire, which stood very high not only as an agricultural and manufacturing county, but as a favourite residence and hunting ground of kings, great nobles, and wealthy gentry. There was no desire then among such people to be near London ; it did not mean anything to them. It was rather an advantage, too, not

The Market Place, Devizes

to be on the sea coast, which was always troubled by pirates and buccaneers. Above all, more than half of Wiltshire was clean dry pasture land, not merely good for grazing and easy to cultivate when required, but easy and pleasant to move about on at a time when the lower parts of England were greatly obstructed by forests, thickets, undrained or half-drained marshes, and when much of the uplands not chalk were wet moors, which they sometimes are still, or hills clad like much of the lowland with tangled wood and bushes. Other counties in the South have considerable areas like the Wiltshire downs, but no other county has its greater half thus composed. So Wiltshire stands preeminently for the county of chalk downs, so much so that many strangers think of it as nothing else. At any rate any one who is able to picture the England of all periods before the great civil war of 1642–5 and who knows Wiltshire, can easily understand why it was till then and for long afterwards so important agriculturally and considered so particularly favourable a county to live in. It is difficult for people of our time, accustomed only to the beautifully-kept appearance of modern England, to realise how rough, wet, and tangled a country most of it was in the middle ages. More than half of Wiltshire, however, was then, as a good deal of it still is, short down turf, or open unobstructed country with little or no wood. This was much more prized by people in those and earlier times than it is now, when nearly the whole of England is, by comparison to the past, as clean as a garden. It was a still more valuable feature in the time of the Ancient Britons, when Britain generally was far rougher and more impassable than even in the middle ages and with its forests yet more infested with wild beasts. The Wiltshire downs were then very much what they are now and were in consequence the most populous part of England, a curious contrast to the present day, when they are among the least populous.

3. Size. Shape. Boundaries.

Wiltshire is very compact in shape and, speaking roughly, is a parallelogram running north and south. It is bounded on the north and north-west by Gloucestershire, on the west by Somerset, on the south and southwest by Dorset, on the south-east by Hampshire, and on

Bradford-on-Avon

the north-east by Berkshire. For a short distance at the latter point the infant Thames forms the county boundary, but at almost all others the boundary is an artificial one, not determined that is to say by a river or a sharp range of hills. Wiltshire, unlike some counties, has no outlying fragments in other shires, though in seven instances parishes are divided by its boundary-line. It may be noticed too as a peculiarity that a majority of its chief towns are situated near its borders. Its capital, Salisbury, is only six miles from Hampshire, while Warminster, Westbury, Trowbridge, Bradford, Malmesbury, Cricklade, and Highworth are still nearer the edge of the county. This is partly due to the fact of so much of the interior being high pastoral upland. The world-famous town of Bath is only three miles outside the borders of Wiltshire, while on the other side the county line runs through part of the Berkshire town of Hungerford and actually divides the great room of the Bear Inn in which William of Orange on his march to London met the notables of the kingdom and came to the agreement which set him on the throne of England.

Wiltshire is fifty miles long from north to south and thirty in width from east to west. It contains 864,105 acres or 1354 square miles. It stands fifteenth in size among English counties and covers about one forty-third part of the area of England and Wales.

4. Surface and General Features.

The surface characteristics of Wiltshire are very interesting and very clearly defined ; more so indeed than in most counties. For, speaking broadly, the county is laid out by nature into two grand divisions, in most physical respects bearing very little resemblance one to the other. The greater portion—about three-fifths of the

whole, speaking broadly again—is pastoral chalk upland
threaded by narrow but luxuriant valleys whose clear
streams and general characteristics are those of a chalk
country.   This region covers roughly the southern and
middle portions of the county.   The remaining two-fifths

Bromham Church

is chiefly comprised by the northern portion, the marked
line of division running in a south-westerly direction
from high up on the eastern border of Wilts near Swindon
to the neighbourhood of Trowbridge and Westbury near
the Somersetshire border.   This slice of the county, which

it would be sufficiently accurate to describe as its north-western portion, is mainly fat lowland throwing up hills of a sort here and there, but hills such as are seen in the midland counties—generally covered, that is to say, with large timber, crops, and luxuriant hedges. The geological characteristics of these two greatly contrasting regions will be alluded to under another heading. But the great fact of Wiltshire scenery and indeed of Wiltshire life is this cleavage into two parts—the breezy, open, thinly-enclosed chalk uplands where the sheep is the chief figure among material interests on the one hand ; the heavily-enclosed low-lying meadows and pastures where the bullock and milch cow hold chief sway on the other.

The upland division consists in the main of two contiguous areas—the larger one being Salisbury Plain with its fringes and bordering downs that are practically part of it, though bearing sometimes other designations ; the smaller and northern one the Marlborough Downs. They are separated by the Pewsey Vale, a long wedge of low country some three to four miles in width. But though a brief break in the continuity of the great Wilt-shire downland and a celebrated farming district, the Pewsey Vale is not on closer acquaintance like the low country of north-west Wilts. In spite of its fine timber and large stretches of level farming land it looks what it really is, a part of upland Wiltshire, not of its lowland. So much of Wiltshire being what is known as a down country, it is not surprising that outsiders believe it all to be of such a character and are apt to picture the county in their mind as one vast Salisbury Plain. Adjoining counties, Berkshire, Hants, and Dorset, have large tracts contiguous to Wiltshire of precisely similar country, but not in so large a proportion to the rest of their area. These counties, together with Sussex and one or two others in less degree, have well-known ranges or tracts of chalk down, but Wiltshire is preeminently the county of

In Old Swindon

downs.   Only a few English counties have a strong dis-tinguishing characteristic that suggests something definite when the name is mentioned.   Cumberland for instance is associated with mountains, Cambridgeshire and one or two neighbours with fens, Herefordshire and Worcester-shire with orchards.   Wiltshire, as I have said, chiefly represents (and rightly so) to the popular imagination the chalk upland and all which that means, namely wide outlooks, vast stretches of springy turf intermingled with far-reaching unfenced fields of grain, roots, or artificial grasses. It means white roads rising and falling like waving lengths of tape over a varied open patchwork of colour, that of green prevailing. It means avenues and rows of fine beeches on the ridges, and belts of water meadow enriched by clear streams winding through the upland, sprinkled with great elm and other timber, and bordered with villages of thatch and flint grouped around their ancient churches. It means, indeed, a great deal more than this. A down country appeals strongly to the affections and imaginations of many people. It is a thing, in fact, quite of itself, for no formation but chalk displays anything like the same type of surface. The Cotswolds, for instance, a far-extending range just outside Wiltshire, are neither in themselves nor in their buildings in the least like it. The northern and western moors are still less so. For these, except where fenced off and cleaned into pastures, are mostly covered with rank growth such as heather, ferns, and poor grasses, and offer frequent obstructions such as rocks and bogs. There is no other natural wild country in England where a man can gallop a horse straight ahead as he may over the short down turf of Wiltshire and its neighbours. Except for a little in France there is nowhere any chalk outside England, and Wiltshire stands as the chief representative of the English down country, which, as I have shown, is not only quite different in appearance from anything else in our islands but from anything else in the world. It may be noted that

many of the principal towns in the county stand along the
borders of upland and lowland Wiltshire, nestling under
or near the steep green wall with which the chalk plateau
in curving, zigzagging course from N.E. to S.W. drops
into the plain.   Swindon, Calne, Devizes, Warminster,
Westbury are all more or less thus situated.   Salisbury

Bemerton Church and Rectory

Plain fills the whole heart of the county, practically
touching Hampshire on the east for some distance.
Salisbury at its southern extremity is enveloped amid
downs, though the Avon flowing thence southward out of
Wiltshire, and reinforced by the Wily and Nadder, cuts
a wide valley which gives a richer and more umbrageous
appearance to this southern gateway of the county.

Salisbury Plain, or "the Plain" as local people call it, is not flat as the name might imply, but a high rolling plateau. It is about twenty miles from north to south, from Pewsey to Old Sarum that is to say, which respectively represent its two limits. Its average width is from twenty to twenty-five miles, either from Westbury to the Collingbournes or from Maiden Bradley to Cholderton. Its south-eastern quarter is now more or less occupied by various military camps, but the greater part of the Plain is unaltered and unspoiled. Its general appearance is billowy, not unlike an ocean in a long swell after a storm, and the round crests of the hills are often tufted with a weather-beaten group of Scotch firs. The larger part is short grass, but a great deal of land was ploughed up in the time of the Napoleonic wars when wheat was high in price, and some more in the 'sixties and 'seventies when the price was again high. There are almost no fences, and the tracts of tillage-land lying on the open sweeps of grass do not greatly detract from its solitary and impressive appearance. Here and there are large homesteads with some protecting timber round them. There are also a few villages, mostly set on the two chief roads that penetrate the Plain. One of these roads runs south along the banks of the Avon (the Christchurch not the Bristol Avon) which cuts its way in a narrow valley through the heart of the Plain from Pewsey and Upavon to Amesbury —a large village that may almost be called the little metropolis of the Plain—and so to Salisbury. As you pass over the open country, however, the snug thatched villages that are grouped along these arteries, with their sometimes fine churches, are often invisible, and the sensation of solitude is almost as great as the old travellers of the eighteenth century describe it. The narrow valley of the Wily, running from near Warminster to Salisbury in a S.E. direction, divides the Plain in the more usually accepted sense of the term, from the triangle roughly formed by that river and the Nadder coming to a point at

The King's House, Salisbury Close

Wilton. These downs are sometimes called the "South Plain" but were formerly not distinguished from the other. A bold range of chalk hills also runs from the S.W. corner of the county near Cranborne Chase to Wilton, overlooking on their southern slope the long Vale of Chalk which is watered by the little river Ebele, one of the most secluded and characteristic of South Wilts valleys. The Wily is perhaps the most beautiful of them all. Cobbett, who rode much about England, tells us that since he first saw it in youth he had carried its memory all over the world. He also speaks of his ride down the Avon valley as the most enjoyable he had ever taken in his life. He was a man who saw poetry not in landscape alone but also in the agricultural or pastoral associations that give it further significance.

All along the high edges of the Plain, where they overlook the valleys or the outside low country, are prehistoric camps, whose escarpments and fosses on the short naked turf are often visible from afar. On the northern escarpment of the Plain the most conspicuous heights are those of Easton and Pewsey, Edington and Bratton, while over the Wily valley Battlesbury and Scratchbury camps are the most prominent. South of the Wily and Nadder valleys White Sheet and Winklebury hills form the outposts of the county.

All over the Plain, particularly about Stonehenge of which we shall speak later, are the tumuli or barrows of the unknown dead of remote ages. In a region like this, overlaid with short sheep-grazed turf, these mounds are always conspicuous and often show themselves clearly defined against the sky-line like pimples on the ridges of the hills. Sheep too will be seen grazing here in large flocks, and not usually scattered all over the hills as in Wales and the North, but accompanied as in olden times by the shepherd with his crook and faithful shaggy bob-tailed dogs. Indeed till the recent multiplication of railroads and development of motor traffic, Salisbury

Plain cut off the northern from the southern part of Wilts so completely that their respective inhabitants saw very little of one another.

The Marlborough Downs, so-called because the town of Marlborough lies almost inside their eastern edge, are a repetition of Salisbury Plain, and about half its extent.

High Street, Marlborough

Their outer escarpments are extremely fine and bold, particularly on the southern side, where they look across the vale of Pewsey to the less abrupt ramparts of the Plain. Martinsell is higher than any summit on the latter, being over 900 feet and having an almost perpendicular face. St Ann's Hill (958 ft.) and Milk Hill (964 ft.) are higher still, though Inkpen Beacon on the east of the Plain and just in Hampshire is just over 1000 feet and is actually the loftiest chalk hill in England. The Kennet and its tributary the Og drain this plateau, which stretches far into Berkshire and contains on its northern slope the famous White Horse made familiar by name to most Englishmen by the author of *Tom Brown's Schooldays*. Within and all around the Marlborough Downs, too, are many camps and innumerable barrows and the sites of battles between the Danes and Saxons, together with those two prehistoric wonders that alone in Britain rival Stonehenge, the tumulus of Silbury and the temple of Avebury. The celebrated London and Bath road traverses the Marlborough Downs, ascending the Kennet valley to its source on its route to Calne, while another main road, heading north from Marlborough to Swindon, runs on their edge or within them, and along both these routes are scattered ancient villages and churches. The outlook over this northern block of downland from Marlborough common is very remarkable. The same clusters of firs, or occasionally of stunted beeches, crown some of the hill tops as on Salisbury Plain, while the same unfenced breadths of tillage lie here and there on the waving ocean of verdure. Wire fencing has been introduced to a small extent of late years on the Wiltshire downs for the grazing of cattle, which spoils the old immemorial appearance of the sheep-bitten turf, while the post fences look out of keeping with the sentiment and tradition of the country. A few years ago a man who knew his bearings could ride on horseback from the north-east corner of the downs near Swindon to the south-west limit of Wiltshire

at Cranborne Chase—nearly sixty miles—and but for
three or four miles in the Pewsey Vale be practically all the
time on down turf and meet with no obstructions.   On
the northern escarpment of the Marlborough Downs
Liddington and Barbury castles and Hackpen Hill are
the chief heights ; on the southern, St Ann's Hill, Milk

Trowbridge, Fore Street

Hill, Huish Hill and Martinsell.   Four out of the five
"white horses," cut large on the chalk slopes which are
such conspicuous objects in Wiltshire, are on the Marlborough Downs, but none of them is more than a century
old.   The only ancient one is that above Westbury on
the northern rampart of Salisbury Plain, but this was
unfortunately recut and bears small resemblance to the

quaint animal it replaced, which was popularly supposed
to have been made in celebration of Alfred's victory over
the Danes. But though this part of the county is somewhat the larger half and is apt to stand for the whole of
it in the minds of strangers, the lower and fatter portion
is in some ways more important. It is of course much
richer in soil, is more populous, and contains more towns
and great country seats. Part of the famous "Vale of
the White Horse" lies about Swindon and Highworth.
This region opens wider as we travel westward owing to
the south-westerly trend of the downs. If we stand on
the brink of the latter above Swindon or Wootton Bassett
and look north over the great stretch of woodland, hedge-
row timber, and green pasture, the distant heights of the
Cotswolds in Oxfordshire and Gloucestershire bound the
view. Along here too, the downland rises from the
plain by two steep steps, the intervening terrace a mile or
more wide and known in geology as the "Lower Chalk,"
carrying farms and villages on its level surface. This
formation is well calculated to surprise the stranger who
approaches the upland from Wootton Bassett, we will
suppose, and after laboriously toiling up the long and
steep white road to the skyline, sees before him an even
longer and steeper road ascending to another and still
loftier plateau.

Standing again on the brink of this down country
above Calne, or on Roundway above Devizes, and looking
north and north-west, the rich low country, though
rising here and there into wooded ridges, has the general
appearance of a wide leafy plain fading insensibly away

into Gloucestershire and Somerset. Within it lie the towns of Highworth, Swindon, Cricklade, Wootton Bassett, Calne, Malmesbury, Melksham, Bradford, Chippenham and Trowbridge, with Warminster and Westbury in its narrow southerly wedge. This is in the main a dairy country, producing milk and cheese. This division

Downton Village and Cross

of Wiltshire has in its appearance no distinct unmistakeable characteristics as has the other, but is a typical English midland or west midland country with here and there, as at Box and Castle Combe on its western border and Cley hill near Warminster, some bits of more broken and striking scenery. It abounds however in picturesque villages, old manor houses, and fine churches.

A special natural feature and one of the glories of
Wiltshire is Savernake forest, which occupies an undu-
lating plateau just south of Marlborough and is about
sixteen miles in circumference. It is the heart of an old
Crown forest of much larger extent, though probably not
all timbered, for the term forest did not then and does not
now of necessity mean woodland, but merely Crown
property originally reserved for hunting, which carried
special laws and rigid customs as affecting the inhabitants,
both for those on its edges and those settled within its
bounds. The present forest of Savernake together with
much of the surrounding country has been the property
of the Bruces (Marquises of Ailesbury) for over two
centuries, and before that had been granted to their
maternal ancestors the Seymours by Henry VIII, who, as
will be remembered, married Jane Seymour for his third
wife. Their ancestors again, the Esturmys, had been
Crown wardens for centuries and the horn by which they
held office still exists at Savernake. The forest is generally regarded as the most beautiful in England. Its
ferny slopes and glades, sprinkled with varied timber and
containing many magnificent and remarkable old oaks, are
in all probability much what they were when kings hunted
through them in the middle ages. But planted about
two centuries ago are many miles of splendid beech
avenues that like stately Gothic aisles penetrate the wilder,
more open tract of the ancient forest. In olden days the
royal forests of Pewsham, Braden, and Chippenham inter-
twined their branches so closely that it was said a squirrel
could leap northwestward from Calne to the bounds of

The Grand Avenue, Savernake Forest

the county. The remains of Pewsham forest may still be seen in the park at Bowood, otherwise it has all gone. What little remains of the famous Cranborne Chase is mostly in Dorsetshire, though the original chase spread far into Wiltshire and was the occasion of bitter disputes and lawlessness and no little bloodshed in the seventeenth and the eighteenth centuries. The term *chase* is equivalent to *forest* but was usually applied when the rights belonged, as in this case, to a great subject and not to the Crown.

5.   Watersheds and Rivers.

The low-lying northern and north-western belt of Wiltshire is watered by two famous rivers, though neither attain much size within the county limits—the Bristol Avon and the Thames itself. A southern spur of the Cotswolds, thrown out southwards near Tetbury, parts their infant streams, sending the one to London and the other to Bristol. But the former great river in its course hugs the county border so close that it is very much more important as a Wiltshire stream and is indeed the chief factor in draining north-west Wilts. It flows with slow and peaceful current, like Shakespeare's Avon beyond the Cotswold, past Malmesbury and Chippenham, just avoiding Trowbridge, to Bradford, whence it soon passes out of the county to Bath. The Newnton brook joins it at Malmesbury and near Chippenham the Marden comes in from its source in the chalk of the Marlborough downs above Calne. No stream worth noting flows into the

Upper Thames but the Cole, which striking northward divides for its last ten miles the counties of Berks and Wilts.

The other rivers of Wiltshire are of a different character and appearance and belong to the chalk system. The Kennet rises near Avebury and flows eastward through Marlborough and Ramsbury, leaving the county at Hungerford, its only tributaries in Wiltshire being the Og from the north, already mentioned, and the Shalbourne brook from the south near Hungerford. It is curious that a county should have its two chief rivers of the same name, particularly when each is respectively typical of the two physically dissimilar portions into which it is divided, though it may be noted that "Avon," being merely the Welsh or Celtic word for river, has naturally become the permanent name of a great many British streams. The Salisbury Avon rises near Savernake forest, and becoming something more than a brook in the Pewsey Vale, receives before leaving it the Marden (not the Calne stream of that name) and then, instead of following the vale towards Devizes, turns southward and cuts its way as already mentioned through Salisbury Plain to the cathedral city. Here it meets the Wily flowing down its own vale from Warminster, already reinforced at Wilton by the Nadder coming up from the west and the Dorset border. At Salisbury, which for this reason is sometimes jestingly called "the sink of the Plain," a smaller stream, the Bourne, flowing from the north-east, joins its waters to the others in the fine expanse of green meadow kept fresh with the stir of so many meeting streams and so finely dominated by the spire of the great cathedral. From Salisbury to Downton these combined rivers flow south in a single channel under the name of the Avon, which after receiving the Ebele from the Vale of Chalk and passing into Hampshire, rolls onward by Fording-bridge, Ringwood, and Christchurch to the sea. While

Wilton House from the River

the northern Avon in appearance, habit, and the fish it produces is like a midland river, all these other streams rising in and flowing over the chalk are as clear as crystal, more particularly the Wily. They ripple oftentimes over gravelly shallows and are famous for their trout and grayling, fish that are fastidious as to the purity of the waters they inhabit. Another curious feature about these chalk streams is the manner in which near their sources and above the strong welling springs that are their main support, they dry up entirely in summer time. The Kennet, for instance, for some miles above the village of that name near which is one of these unfailing springs, often becomes a perfectly dry ditch for many months. From this cause two villages situated upon it derive their names of Winterbourne (i.e. a stream only flowing in winter). There are many Winterbournes in Wiltshire, three on the little river Bourne alone near Salisbury, and named for the same reason. The upper part of the Og, too, for some miles presents the same spectacle for part of the year. There are no natural lakes or meres of any consequence in Wiltshire, though there are several reservoirs and ornamental sheets of water of considerable size.

6. Geology and Soil.

By Geology we mean the study of the rocks, and we must at the outset explain that the term *rock* is used by the geologist without any reference to the hardness or compactness of the material to which the name is applied; thus he speaks of loose sand as a rock equally with a hard substance like granite.

Rocks are of two kinds, (1) those laid down mostly under water, (2) those due to the action of fire.

The first kind may be compared to sheets of paper one over the other. These sheets are called *beds*, and such beds are usually formed of sand (often containing pebbles), mud or clay, and limestone or mixtures of these materials. They are laid down as flat or nearly flat sheets, but may afterwards be tilted as the result of movement of the earth's crust, just as you may tilt sheets of paper, folding them into arches and troughs, by pressing them at either end. Again, we may find the tops of the folds so produced wasted away as the result of the wearing action of rivers, glaciers, and sea-waves upon them, as you might cut off the tops of the folds of the paper with a pair of shears. This has happened with the ancient beds forming parts of the earth's crust, and we therefore often find them tilted, with the upper parts removed.

The other kinds of rocks are known as igneous rocks, which have been melted under the action of heat and become solid on cooling. When in the molten state they have been poured out at the surface as the lava of volcanoes, or have been forced into other rocks and cooled in the cracks and other places of weakness. Much material is also thrown out of volcanoes as volcanic ash and dust, and is piled up on the sides of the volcano. Such ashy material may be arranged in beds, so that it partakes to some extent of the qualities of the two great rock groups.

The production of beds is of great importance to geologists, for by means of these beds we can classify the rocks according to age. If we take two sheets of paper, and lay one on the top of the other on a table, the upper one has been laid down after the other. Similarly with two beds, the upper is also the newer, and the newer will remain on the top after earth-movements, save in very exceptional cases which need not be regarded by us here, and for general purposes we may regard any bed or set of beds resting on any other in our own country as being the newer bed or set.

The movements which affect beds may occur at different times. One set of beds may be laid down flat, then thrown into folds by movement, the tops of the beds worn off, and another set of beds laid down upon the worn surface of the older beds, the edges of which will abut against the oldest of the new set of flatly deposited beds, which latter may in turn undergo disturbance and renewal of their upper portions.

Again, after the formation of the beds many changes may occur in them. They may become hardened, pebble-beds being changed into conglomerates, sands into sand-stones, muds and clays into mudstones and shales, soft deposits of lime into limestone, and loose volcanic ashes into exceedingly hard rocks. They may also become cracked, and the cracks are often very regular, running in two directions at right angles one to the other. Such cracks are known as *joints*, and the joints are very important in affecting the physical geography of a district. Then, as the result of great pressure applied sideways, the rocks may be so changed that they can be split into thin slabs, which usually, though not necessarily, split along planes standing at high angles to the horizontal. Rocks affected in this way are known as *slates*.

If we could flatten out all the beds of England, and arrange them one over the other and bore a shaft through

	NAMES OF SYSTEMS		CHARACTERS OF ROCKS
TERTIARY	Recent & Pleistocene		
	Pliocene		sands, superficial deposits
	Eocene		clays and sands chiefly
SECONDARY	Cretaceous		chalk at top sandstones, mud and clays below
	Jurassic		shales, sandstones and oolitic limestones
	Triassic		red sandstones and marls, gypsum and salt
PRIMARY	Permian		red sandstones & magnesian limestone
	Carboniferous		sandstones, shales and coals at top sandstones in middle limestone and shales below
	Devonian		red sandstones, shales, slates and limestones
	Silurian		sandstones and shales thin limestones
	Ordovician		shales, slates, sandstones and thin limestones
	Cambrian		slates and sandstones
	Pre-Cambrian		sandstones, slates and volcanic rocks

them, we should see them on the sides of the shaft, the
newest appearing at the top and the oldest at the bottom,
as shown in the figure.   Such a shaft would have a depth
of between 10,000 and 20,000 feet.   The strata beds are
divided into three great groups called Primary or Palaeozoic,
Secondary or Mesozoic, and Tertiary or Cainozoic, and the
lowest of the Primary rocks are the oldest rocks of Britain,
which form as it were the foundation stones on which
the other rocks rest.   These. may be spoken of as the
Precambrian rocks.   The three great groups are divided
into minor divisions known as systems.   The names of
these systems are arranged in order in the figure with
a very rough indication of their relative importance,
though the divisions above the Eocene are made too
thick, as otherwise they would hardly show in the figure.
On the right hand side, the general characters of the rocks
of each system are stated.

With these preliminary remarks we may now proceed
to a brief account of the geology of the county.

The line that divides Wiltshire physically into two
portions so different in appearance it is hardly necessary
to state corresponds with the geological division which is
the cause of this difference.   This line may be described
with sufficient precision as starting at Bishopstone on the
Berkshire or north-east border, and running south-west
by Wroughton, Compton Bassett and Cherhill, Heddington, Devizes, Erlestoke, Westbury, and Warminster to
Maiden Bradley.   To the south and east of this, as we
have seen, is the country of the chalk downs and the
valleys which intersect them.   That on the north-west

and north is as different in geological formation as it is in general aspect, a fact due to the presence of the clays and rocks of the oolite formations. Everywhere the dividing line between these two great rock systems, the Cretaceous and the Oolitic, is clearly marked by the steep and abrupt escarpment of the chalk such as forms the chief feature of

Castle Combe

the landscape on the Great Western Railway on both sides of Swindon, beyond Wootton Bassett, and again at Devizes, Westbury, and Warminster. Here and there on this great expanse of downland the tops of the hills retain some remains of the later Tertiary strata which once covered the whole chalk district in the shape of beds of gravel, sand, or clay. Where this is the case the open down gives place to thick woodlands, as at Savernake forest on its patch of Tertiary sand and clay, at the west woods on the Marlborough downs, and at Grovely where are the Great Ridge woods which for several miles along the high down above the Wily form the almost continuous and ancient forest of Grovely. Two comparatively wide valleys intersect the chalk area, the vale of Pewsey in the north and that of Wardour in the south. Here the beds of the Greensand which underlie the chalk come to the surface, affording abundant water and a fertile soil which if left to itself would everywhere be thickly clad with wood. The beds of Lower Greensand at Seend contain a rich ironstone which has been freely worked in the past and is still utilised to a trifling extent.

The small patch of wooded and heathy country in the extreme south-east corner of the county owes its character to the Tertiary sands which there cover the surface, and really forms a part of the New Forest district of Hampshire. With this exception, and that of the Tertiary sands and drift already mentioned as occurring here and there on the highest parts of the chalk downs, and of the later river gravels found along the course of the Bristol Avon in the north of the county and in the neighbourhood of Salisbury in the south, the formations of a later age than the chalk which covers such wide areas to the eastward of the county are hardly found in Wiltshire. The most remarkable feature however on the downs of north Wilts and the adjacent fringe of Berkshire, which gathers still greater interest from the fact that it does not occur on the precisely similar uplands of

Salisbury Plain, is the "Sarsen" stone. These great grey boulders lying thickly strewn in places on a chalk soil, so utterly dissociated even to the most unskilled eye from anything like surface rock, have an almost uncanny appearance. The origin of the word Sarsen is unknown. It is sometimes suggested that it may be a corruption of Saracen, as that term in the early middle ages was freely

Ramsbury

applied to all foreigners from the east or those that had anything mysterious about them. On these chalk uplands the Sarsen was most unmistakeably a mysterious stranger to a generation without any geological science to explain what in any case is a great curiosity. They vary in size from that of a bucket or less to masses weighing fifty tons or more. In places they lie so thickly strewn along some narrow winding trough in the downs for a mile or two as to give the impression of a flock of sheep feeding, which earned them as long ago as we have any record the local name of "Grey Wethers." They are covered with lichen, and some of the lichens growing on them are not found any nearer than the granite tors of

Church House, Potterne

Dartmoor and Cornwall. They are still numerous in the neighbourhood of Marlborough, the "Grey Wethers" of Piggle dene and Lockeridge dene having recently called for the intervention of the public by partial purchase of both, as they were doomed to wholesale extinction in the building-stone market by the owner. But once upon a time these stones were far more numerous than now, north that is of Salisbury Plain, upon which there is no evidence in buildings or elsewhere of any number having ever existed. For in the last three centuries these stones have been used in wholesale fashion for local building purposes: villages, farm houses, barns, "pitched" paths, and gate posts having been in whole or part created out of them, for the boulders could be easily split by simple processes into rectangular blocks. That they formed the sole material of the crude temples and tombs of prehistoric man in these regions, as we shall see later, entitles them to some special notice.

They are composed of a hard and durable sandstone displaying a sugar-like structure when broken, and are believed to be the harder nodules of a stratum of the Tertiary sand which once covered the Chalk of north Wilts. The softer portions of this sand were washed away and these hard nodules settled down on the surface of the chalk where they still lie. The holes sometimes seen in them are believed to be the marks left by the roots of palm trees which grew in the sand during the tertiary period when our climate was much warmer than now. The outcrop of the Upper and Lower Greensands and of the thin line of Gault clay which separates them is generally to be followed in the chalk districts by the line of villages situated upon it. The rainfall on the downs gradually percolating through the Chalk and Upper Greensand is held up by the clays of the Gault below and breaks out on the sides of the hills and in the valleys. This accounts for the fact that the villages are situated always on the edge of the chalk downs, or in the valleys, like those of the Wily and Salisbury Avon, and never on the downs themselves where water would be scarce.

Of the rocks which lie below the Chalk, the Purbeck and Portland beds are found at Swindon, covering only the top of the hill on which the old town stands, and in the south of the county over a much larger area in the neighbourhood of Tisbury, Fonthill, and Chilmark. In both these localities the Portland stone has been largely worked for building purposes. The quarries at Swindon are for the most part exhausted, but those at Chilmark still produce large quantities of freestone. It was from these that the stone came of which Salisbury Cathedral is built.

Below the freestone and sands of the Portland beds comes the Kimmeridge clay, forming the flat pasture-lands which lie at a little distance from the foot of the chalk escarpment near Swindon, Wootton Bassett, and Calne. The Kimmeridge clay is largely worked for brick-making at Swindon and, in the process of digging, numbers of bones of gigantic saurians or fish-lizards, which once inhabited the swampy estuary in which this clay was deposited, are found. Beyond these flat lands in the north of the county rises a line of hills formed by the Coral Rag, extending from Wootton Bassett by Hillmarton, Bremhill, and Calne, where the various fossils and shells discovered show that a coral reef once existed. Westward of the drier and higher lands of the Coral Rag lies the belt of the formation immediately below it, the Oxford clay, on which the low-lying grass land about Melksham and Chippenham are situated. Still further west and north lie the lower beds of the oolites, giving place to the Lias formation in Gloucestershire and on the western borders of Wilts, and in the hilly country about Bradford-on-Avon, Monkton Farleigh, Corsham, and Box to the Great Oolite or Bath stone, which is largely quarried in underground galleries throughout that district. Great

Flemish Houses, Corsham

quantities of this, one of the most durable, easily worked, and best known freestones, are used over all the south of England for building purposes. This was the stone used for most of the ashlar or facing work in the churches of north Wilts in mediaeval times, and indeed long before this by the Romans in their villas and baths.

7. Natural History.

It will be well perhaps before glancing at local conditions to remind the reader that existing submarine remains and geological science have virtually proved that at no great distance of time, as geology counts time, the British islands were part of the Continent of Europe. At some period subsequent to this it is equally certain that the land was entirely submerged and had in consequence, when the ocean again receded, to be once more stocked anew with animal and plant life. This was done from the neighbouring unsubmerged mainland to the south and east. There is further evidence that this second period of union with the Continent was of no great length in a geological sense, not long enough at any rate for these islands to assimilate such full measure of the fauna and flora of the mainland as they might have been capable of naturalising. Following the same line of reasoning it is natural that those parts nearest to the Continent should be richer in such occupants, both animal and vegetable, than those more distant from it ; the south of England for instance than the north, and more particularly the island of Britain than that of Ireland, while both were less amply stocked than France or Belgium. But this must be understood, of course, as applying to the times long before history, and unknown centuries before the steady increase of population and the changes wrought by the more artificial mode of existence that has grown up in the last two or three hundred years. The habits, customs, and laws of different nations, particularly as regards land, have in all their relations inevitably affected all animal and plant life. As affecting natural history Great Britain and Ireland have been as one country, the social condition of which has been peculiarly favourable to the preservation of wild life in most forms. Practically the whole of Great Britain has been con-

Old Cross, Aldbourne

tinuously protected by private ownership for purposes of sport, and by other laws of trespass, not only against the exterminator, but to a great extent even against the mere disturber of wild life. Though these precautions have existed chiefly for the maintenance of sport and in the same interest have even waged war on certain species of birds and animals, in the main they have made of Britain a vast preserve, in which bird and plant life not greatly regarded by the ordinary man have flourished in security.

The hedgerows of England—which it must be remembered are unique and exist in no other country in the world to any extent worth mentioning, filling the eye of the foreign visitor with surprise and admiration—

Avebury Church

are an immense protection to the wild life, animal or vegetable, of this country. The consequence is that whatever comparison there may be in a mere list of specimens, Great Britain as a rich haunt at once of birds, wild flowers, and the smaller animals presents an unrivalled field. Countries like France and Belgium have particular tracts of country where wild life is protected, but the land is for the most part cultivated by small owners, who for a century or more have waged a ruthless war against every wild living thing, bird or beast, with lamentable success. Wiltshire, with its two distinct divisions, is in its wild life well up to the average of a southern county. On its downlands there are more hares than in any other part of England nowadays. Foxes and rabbits of course abound, while of less common animals the badger, of shy nocturnal habits and rarely seen by day, is fairly numerous in the chalk escarpments of the downs with their sometimes hanging woods. The otter is found on most of the rivers. The polecat is practically extinct, but stoats and weasels still abound and squirrels are plentiful, particularly in Savernake forest.

The character of a country—woodland, plain, fen or whatever it may be—mainly determines the character of its bird life. For the bustard, the noblest of all game birds, wide open spaces are necessary, and hence the "brecklands" of Norfolk and the plains of Wiltshire were the main haunts of these birds when they were a permanent feature of our land. Salisbury Plain was famous for its bustards till the nineteenth century, during the early part of which they rapidly became extinct. The last seen was a flock of seven that visited the Plain in 1871, probably from Spain. Herons are fairly common and there are heronries at Highworth, Bowood, Savernake, Longleat, Fonthill, Compton Park, and Longford Castle. The peregrine falcon still survives and a pair bred some years ago in the spire of Salisbury Cathedral. As there

are no tidal rivers in Wiltshire nothing need be said of
the birds, waders and others, that generally frequent them.
Gulls of course are frequent visitors, sandpipers, though
sometimes seen, prefer to breed up the more rapid streams
of the north and west. The writer has seen a pair of
the lesser tern on the Avon near Pewsey in June. In
hard winters great flocks of wildfowl come up the Avon

The White Horse, Westbury

to Salisbury and ascend the various streams, particularly
the Wily.

The dotterel used to breed regularly on the downs
but of late years has become a very rare bird. The
raven till recently nested in several well-known haunts
but is now a doubtful resident. The nightingale abounds
near Salisbury but in other districts only where oak

woods exist. The reed warbler is generally distributed, while the Dartford warbler is found among the gorse brakes on the downs, the hawfinch nests regularly in Savernake forest and elsewhere, and the kingfisher still haunts the streams, though it is to be feared in decreasing numbers.

Save in essentially wooded districts where the oak excels, the elm is the finest and most prevalent Wiltshire tree, and in the hedgerows of north-west Wilts as well as in the valleys of the chalk streams shows to immense advantage. Beeches do well on the chalk and attain a very great size.

In plant life the county has a very wide and varied flora which includes many rare and remarkable species. The lizard orchis (*Orchis hircina*) has been recently found in two places, while the tuberous-rooted thistle (*Cnicus tuberosus*) is particularly identified with north-west Wilts.

The flora of the chalk downs is entirely different from that of the grass lands, oolitic clays, and rocks, while the flowers of the sandy lands in central and southern Wilts are again quite different from those either of the chalk or the clay, so the botanical student living near the line of the chalk escarpment has a wide field. Wiltshire, it is true, has no sea-shore nor mountain plants and only a very insignificant area in the south-east of bog and heath with their peculiar products. But the down plants have a beauty and charm of their own, while the great variety of soils in central and west Wilts affords a corresponding variety of plants growing within reach of a given centre, and of course favours a like variety and profusion of moth and insect life.

8. Climate and Rainfall.

The climate of Great Britain, though often abused for its dampness, changeableness, and lack of sunshine, as compared with that of most other countries, is remarkably equable. Persons who have not been out of England do not know what extreme heat, such as is experienced by most countries, and extreme cold, such as is experienced by very many, mean. Once every three or four years perhaps, for a single day, the thermometer in England touches or exceeds 90° in the shade and throws the whole country into a state of excitement. But this may be called the mere threshold of what is termed hot weather in most of our over-sea dominions and dependencies, as well as over a considerable part of Europe and the United States. English people at home may often feel great discomfort in hot weather; but when they spend their first hot season in any one of these countries, they then realise that they had never before felt the full power of the sun or what it means to be continually subjected to this fiercer class of heat. They also realise how, apart from its trying effect upon the person, it affects matters of daily domestic life, such as provisions, and even clothing and furniture and the prevalence of troublesome insect life. We in England have never seriously to fight the sun, nor is there ever a summer day in which an ordinarily healthy individual cannot follow any outdoor business or recreation with impunity. The same remarks may apply to the opposite of extreme cold,

B. W. 4

of which again in the matter of actual temperature we have
no experience. Both modified heat and cold, it is true,
cause more relative discomfort in a moist island climate
such as Great Britain. But this after all is quite a
different thing and does not demand the precautions that
extremes require. The British climate is changeable and

Parton Church

often called treacherous, but such violent changes as sometimes occur in southern Europe, or in the United States
and Western Canada, where a rise or fall of 80 degrees in
24 hours is not unknown, would be here impossible. The
east winds common to northern and central Europe are
our one serious climatic foe, so far as man is concerned.
They are uncomfortable to most people and insidiously

dangerous to the less robust, but in no other way affect our lives and habits.

Custom means a great deal and even in England there is space for difference of opinion. Most Northumbrians, for instance, will feel a summer day to be uncomfortably hot that a Wiltshireman would only regard as pleasantly genial. Even in Wiltshire the climate differs. On the chalk uplands it is drier and colder, the atmosphere is generally more bracing, though the east winds of spring are proportionately more trying than in the better-protected lower country of the north and west. Snowstorms of notorious severity sometimes occur on Salisbury Plain and the Marlborough Downs, and in the old coaching days these portions of Wiltshire were the particular dread of drivers and passengers travelling between London and the west of England. The extremes of temperature are always greater on a dry soil, and as the Down countries, both for that reason and their greater exposure, are colder in winter, so they are often hotter in summer than the lower country, though more apt to be tempered by a breeze. Other and more local influences, too, affect climate. Districts with a southern slope, for instance, get much more warmed by the sun than those facing the north, while such as are sheltered by hills or downs from the north and east winds are more genial and produce earlier crops. In the year 1906 the average rainfall over the county of Wilts was 31·64 inches and rain fell upon 181 days. These measurements are taken not only in every county in England but at many different stations in each county, and form the subject of valuable reports

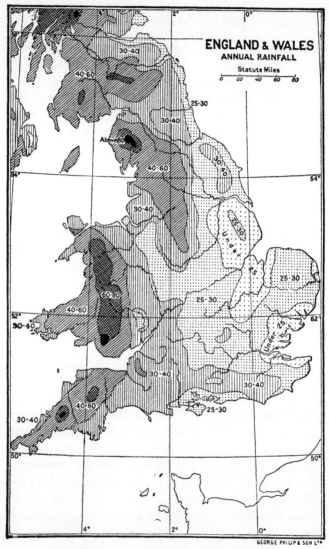

ENGLAND & WALES
ANNUAL RAINFALL

(The figures give the annual rainfall in inches.)

issued by a central Government Bureau. In Wiltshire there are sixteen such gauges which contribute to this general report. The highest rainfall, for instance, in the year given, which was a typical one, was 38·27 inches at Tisbury, the lowest at Swindon, 27 inches. At Shrewton, in the middle of Salisbury Plain, 35 inches of rain fell,

Castle Combe

but it rained there on 188 days whereas at Tisbury it rained on only 160. At Warminster, again, though 38 inches were recorded for the year, only 155 days are credited with the amount of rain (0·005) that *officially* constitutes a rainy one. At Stourhead, close to War-minster, on the other hand, it rained on 200 days, while the measured fall for the year was only 33 inches. The average annual fall of Great Britain runs from 27 to 31 inches, and the rainfall decreases, generally speaking, as you travel from west to east. The mean rainfall of Wiltshire, together with the number of days on which rain falls, will be found closely to correspond, if an average be taken over a term of years, with the average mean of Great Britain. As an illustration of extremes, however, within the limits of England and Wales alone, there are spots in Kent, Bedfordshire and elsewhere in the south-east where the average rainfall is below 20 inches, while in the Snowdon district of Wales and the neighbourhood of Scafell in Cumberland from 150 to 180 inches are frequently recorded as the year's fall.

9.   People—Race, Dialect, Population.

We only know for certain that two distinct types of those prehistoric people whom we call for convenience the Ancient Britons, successively occupied Wiltshire in common with other parts of England. The first, as proved by their numerous remains, were a long-skulled, low-statured race and flourished in the Stone Age, or previous to about 1200 B.C. They were conquered by a taller, round-headed race of invaders who knew the use of bronze. It is thought by many, however, that the Wiltshire downs were largely occupied on the arrival of the Romans by the Belgae, a Teutonic people of more or less the same stock as the Saxons, and comparatively recent immigrants, who within a few hundred years had driven back the previous inhabitants. But with such conjectures we must not concern ourselves here. Enough has been said to show that the Romans, with their stations at Old Sarum, Marlborough, and elsewhere, their many roads through the county, and their villas stationed thereon, held it firmly in the grip of their civilization and such of it as they imposed upon the natives. Of the Saxons too we have

Longford Castle

seen, in noticing the history of the county, how coming up chiefly from the coasts of Hampshire and Dorset, they gradually drove out or subjected the Romanised Britons and made Wiltshire an important province of the kingdom, with Wilton the royal capital of Wessex inside its borders. Modern fiction is responsible for spreading the mischievous fallacy among people careless or ignorant of history that

Wessex meant Dorsetshire, instead of as in actual fact the whole great region from Berkshire to Celtic Cornwall. Wiltshire folk themselves are hardly likely to have been thus misled. Still it is well to remind ourselves how great a part our Saxon population played in the kingdom of Wessex. There is beyond doubt, however, as in most parts of England, a great deal of ancient British blood among the people of Wiltshire. The old vague notion that the Britons were driven wholesale to Wales and Cornwall is now utterly rejected. It was useful perhaps as a half truth for children on which to build later the truer story that a mixture of knowledge and common sense teaches. The broad principle is now generally held that the further to the west the greater the proportion of British or Celtic blood. There is little doubt that the Saxon invaders were extremely bloodthirsty and merciless, slaughtered the natives freely, and expelled them as regards their possession of the soil. But as the Saxons required labour and slaves it is quite incredible that they should not have retained great numbers of the lower sort as such. But even this is not vital to the question, for of the British women at least, after the inevitable custom of those days, they must have kept still more as wives, since among tribes fighting their way in a foreign country the men would be in a great majority over the women. Among the Britons too it may be remembered that after four hundred years of occupation by Roman garrisons there was of necessity much foreign blood. Evidence survives in inscribed stones showing that regulations for recognised and unrecognised marriages of soldiers existed

in the British-Roman army. For the Roman soldier
settled down in his British garrison, if not for life, for
a very long period. More foreign blood, Gallic, Spanish,
and Italian, came into Britain before the Saxon invasion
than is apparent till we come to investigate even the little
that the stones tell of the story of any one great Roman

Coate Reservoir, Swindon

station and realise what an influence for perhaps three
centuries it must have been. And Bath (Aquae Solis), it
must be remembered, is on the verge of Wiltshire, while
small garrisons and villas were sprinkled about the county.
But Wilts was not exposed to these alien influences to
anything like such an extent as some other districts.

There is some speculation as to the former relationship

between the Saxon tribes who inhabited Wiltshire. Some think that the peoples of north-east Wilts had more affinity with those of Berks and Oxfordshire than with the rest of the county, but these conjectures again are outside our province here. The Danes did not obtain any permanent footing in Wiltshire. The Norman invasion brought very little fresh blood into the English people, great influence though it had in other ways on the life of the country. The Normans merely supplanted the Saxon aristocracy—a new variety, proving in most ways harder and more exacting masters, besides bearing the further odium of being foreigners and usurpers. Normans of a lower degree were scarcely numerous enough to infuse any considerable strain of fresh blood. In the fourteenth century and afterwards large numbers of Flemings were introduced into Wiltshire to start and carry on the business of cloth-weaving. These naturally settled on the rivers for the sake of the water-power, such as the Northern Avon, at Bradford, Trowbridge, Melksham and Chippenham, as well as at Calne, Westbury, and a few other places mostly in north-west Wilts. They inter-married with the people, and their descendants became as other Wiltshiremen. Since then there has been no migration of other races or peoples into the county. Its inhabitants have nothing particular to distinguish them in appearance or physique from those of most southern counties where the Saxon stock is the prevalent one. They have no reputation for stature nor on the other hand for any lack of it. A healthy and hardy race, the labouring class survived some generations of lower pay than almost any county without physical deterioration, though constant inter-marriage in some of the remote villages when it was difficult to move about much is sometimes said to have produced bad results.

The natural old dialect and pronunciation of Wilt-shiremen is a good honest broad Saxon. It belongs to

St John's Church, Devizes

a class of speech that beginning in Sussex extends west and north-west, covering Hampshire, Berkshire, South Oxford-shire, Wiltshire, parts of Somerset, Gloucester, and Dorset, with parts of Worcester and Hereford. It avoids the utterly different and detestable cockney speech which may be heard even in rural Essex. The other South Saxon tongue, if we may call it so, of which Wiltshire speaks a variety, skirts the fringe of the midlands and continues on till it runs into the Welsh-English of the Border to the north of the Bristol Channel and into the Devonian type of English (which again is of another family) at points in Somerset and Dorset. This of course is but generalising, as the border-line of dialects, tone, and manner of

The Moot, Downton

speech, where these meet one another, is broad and vague, though sufficiently accurate for our purpose. The characteristics of this speech are strongest perhaps in Wiltshire, which in a manner is its centre. The initiated can subdivide these counties again into further varieties, though a tongue does not of course necessarily follow a county line unless the latter consists of mountains or wide waters.

By dialect one means, strictly speaking, the use of certain words and phrases peculiar to a district. But the pronunciation of familiar words and vowels, the pitch and tone of voice, are quite as characteristic.

The qualities of Wiltshire speech are slowness and deliberation, sometimes called a drawl, and a broad utterance of its vowels, while the letter *r* is very much hung upon in a soft burr, i.e. with the point of the tongue turned inwards towards the roof of the mouth. The letter *s* in old Wiltshire is pronounced as *z*, the *f* as *v*. These are but one or two leading examples out of many that might be given. They are also characteristic more or less of the whole, not only of the South Saxon speech, but of the Devonian too. But then it is the manner in which all these various sounds are uttered and how they fall on the ear in the various districts that makes the difference, and this cannot be expressed in words. A musical notation would be almost more serviceable, as there is much in the tune and pitch, if one may use the terms, to which different regions have set their speech. A north country-man going through the rural districts from, say, Hastings to Cheltenham, or to Dorchester, and meeting all the way with those who spoke the real local speech, would probably recognise very little difference between them. But any one familiar with Wiltshire speech would recognise it in a moment from Dorset or Sussex by its tones alone. Speaking of " tune " in dialect the Wiltshire form of speech without serious alteration crosses the Cotswolds, but in the low country about Cheltenham and Tewkesbury the last syllable of every sentence springs suddenly to a higher note, in a way unmistakeable to the most careless ear. This is the first faint sign of the Welsh "lilt." Crossing the Severn and proceeding half-way across Monmouthshire or Herefordshire, the broad, soft Saxon still continues in a modified form but drops into the regular

Maud Heath Monument near Chippenham

Welsh sing-song, producing the correct inflexions, that is to say, of a long-forgotten Celtic tongue grafted on to another and Saxon one, a character either inherited from remote Welsh-speaking ancestors or acquired from neighbours who inherited it. Even to touch upon old Wiltshire words would be impossible here. It may be noted however that the old Saxon plural, as *housen* for *houses*, is still occasionally heard, and that Wiltshire folk are generally twitted with a rather wild use of pronouns, as for instance " her hit I but us didn't touch she." One other really curious fact relates to the letter *h*, which has apparently never been aspirated in local speech throughout the south generally and indeed throughout the larger part of England. But this is a habit of comparatively recent growth in Wiltshire. Three or four generations ago the Wiltshireman however broad his tongue seems at least to have aspirated his aitches.

In former days every one in Wiltshire, except perhaps the very great people who went to Court or those who were much about the Universities, spoke the Wiltshire dialect in varying degrees ; squires, parsons, farmers and labourers. Education has long decided that we must now all speak alike or try to do so. But this does not mean that the fine old tongue and its pronunciation is not a matter of great interest, or that it was a bad tongue of which to be ashamed. The old words are quickly dying out. Many modern imported slang words not half so good have crept into even polite English. There is a danger, too, of the terrible and vulgar cockney accent superseding the old varied provincial tones and methods of enunciating even good and grammatical English. The north of England is probably safer in this respect, but the south and west are in much greater peril as London and its influence spreads.

The population of Wiltshire, which is very stationary, was at the last census 271,394, which gives 201 persons to the square mile. As compared with this Lancashire has

2347, Durham 1171, Leicestershire 526, and Radnor only 50, while the average of England and Wales to the square mile is 558. This comparative sparseness of population is due to the fact of Wiltshire being mainly an agricultural and pastoral county and having few towns of any size. This last accounts too for its death-rate being only 18 per 1000 whereas the average of England is 21 per 1000.

10. Agriculture—Cultivations, Stock.

Agriculture, in the wide sense of the word, is by far the most important industry in Wiltshire ; sheep-farming in the middle and south, dairying in all its branches in the north and north-west, being respectively the prevailing forms of it. Grain was grown formerly over a much larger area in both divisions of the county than it is now. Since competition from other countries and our colonies possessing cheap virgin soils permanently reduced its price some twenty-five years ago, there has been a steady decline in production, which has now, however, in all probability reached its lowest limit.

Wiltshire covers an area of 864,105 acres. Of this in round numbers—for the odd hundreds given in the returns for 1907 are of no significance, since every year varies slightly—407,000 acres, or less than one half, produce crops of some kind or hay. The latter accounts for 216,000 acres, or more than half of the crop-producing land. Of this hay crop 163,000 acres come from per-

manent meadow and the rest from the grasses, clover,
and sainfoin grown in the ordinary farming rotation.

There are 47,000 acres in wheat, 22,000 in barley,
54,000 in oats, and 46,000 in turnips, swedes, and man-
golds. The remaining 7000 acres or thereabouts are under
beans, peas, potatoes, and minor products. The larger

The Poultry Cross, Salisbury

half of the grass—the remaining 259,000 acres—as one
would expect in Wiltshire, is in permanent pasture,
including the great down uplands and that proportion of
the low country enclosures not cut for hay or standing in
woodland. Wiltshire still stands tenth among the counties in its production of wheat, fifteenth in oats, and

eighteenth in barley. Its average yield of wheat over ten years is 32 bushels to the acre, in which it also stands tenth, Northumberland being first and Devonshire last of the counties. There are some 500,000 sheep, Wiltshire ranking eighth in this respect among the counties and about even with Somerset and Breconshire. Of cattle there are 115,000, of horses 23,000, and of pigs 51,000.

The principal dairying districts are in the vale of Swindon and westward to the regions round Melksham, Chippenham, and Trowbridge. Till a quarter of a century ago North Wilts cheese was a well-known product of this district. Now however a London caterer would not know even the name, for it has quite disappeared. What cheese is now made is of the "double Gloucester" variety. But the London milk-market with improved methods of carriage is found more generally profitable to the north Wilts farmer. There is no breed of cattle peculiar to Wiltshire, but the shorthorn is by far the most popular type. On the chalk soils the "Hampshire Down" is the most common sheep, and the old fashioned shepherds with crook, weather-beaten cloak, and shaggy dogs may still be seen on the Marlborough Downs and Salisbury Plain following their flocks. For, on account of the unfenced tillage land on these uplands, sheep cannot often be turned out indiscriminately to graze as in Wales and the north of England. In the lower country the heavier "Oxford Down" sheep is the prevailing breed. But crossing all over England is so much in fashion now that it is not safe to attempt the precision in describing these matters which would have been easy enough in former years.

A Wiltshire Farm-yard: Milking-time

The pure Cotswold sheep, for instance, except flocks for breeding, has practically vanished from the Cotswolds. The lower price of wool in these days and the recently developed aversion from big and fat mutton by the British public has dethroned the large long-woolled breeds from their once proud position and compelled them to mate with smaller sheep once held by farmers in less esteem. The Sussex or genuine Southdown, the Dorset, Welsh, and even the Exmoor sheep may also be seen occasionally in the chalk districts of Wiltshire.

Neither hops nor fruit nor any of the smaller products have any place worth mentioning in Wiltshire agriculture. The county is, however, famous for its bacon, a reputation largely due to the factories long established at Calne. In the lower and dairying districts the farms are of all sizes but, generally speaking, of small or moderate acreage. In the down countries however they are usually very large and occupied by tenants operating upon a large scale. The presence of the military and the government owner-ship of Salisbury Plain east of the Avon is a new feature in Wiltshire life. The farmers holding under the Crown, though they may be far away from the sight and sound of the great permanent brick barracks and temporary camps that have transformed the appearance of the once solitary uplands of Tidworth and Bulford, are under agreements to shift their stock at certain times and facilitate in other ways the requirements of military operations.

The turf of the chalk uplands is of especial quality, acquired partly by age and grazing, and when broken up and laid again to grass it requires some thirty or forty

years to recover its former value for sheep. When grain
was high in former days a good deal of this down turf
was broken up for wheat—a proceeding which, when

Wiltshire Bacon, The Ancient Custom of
Burning the Pig

prices collapsed and it had to be put to grass again,
led to much repentance and regret.

One business is carried on more generally in Wiltshire
than in any county in England and that is the breeding

and training of race-horses. There are a very large number of racing establishments, either in or near the downs, and on many of these quiet and remote stretches some of the most famous horses in the history of the English Turf have taken their earliest gallops.

The Town Hall, Wootton Bassett

11. Industries and Manufactures.

There is not a great deal to be said under the above heading as regards Wiltshire. The great cloth-weaving industry of former days is now unimportant. In the middle ages the chief guilds of the county were associated with this cloth trade. It has now been gradually reduced to a minor interest in a few towns and is still declining.

Trowbridge is to-day the chief centre of the business such as it is, though it just exists in a few other places. Bradford, once famous as a wool town, has recently closed its last mill. Every town in the county except Swindon depends mainly upon agriculture in some form or other. The flourishing bacon factories of Calne, and the milk

Mason's Yard; the Corsham Worked Stone Co.

factories or creameries of some other places, have to a limited extent taken the place of the old wool trade once carried on. Melksham does some engineering and so does Chippenham. At Wilton were woven the first carpets in England and a well-known factory there still continues to produce them. Malting and brewing are carried on upon a considerable scale in several towns, while all the minor industries incidental to local life, but not of any consequence for export out of the county, are practised. Swindon, both in its story, its appearance, and its life, offers a complete contrast to every other Wiltshire town, having spread within the memory of middle-aged people from the original little hill-top town of about 3000 souls, now known as Old Swindon, over the mile or so of meadow that formerly divided it from the well-known station on the Great Western Railway. There are some 50,000 people in modern Swindon, nearly all dependent in various ways on the great railroad workshops. For here the Great Western have their locomotive and carriage works for supplying nearly the whole of their system. The works are divided into three departments, one for manufacturing locomotives, another for carriages, and thirdly the rail mill. Swindon is in fact a town of mechanics and engineers, with churches, lecture halls, and a great number of advantages which are secured partly by cooperation and partly at the expense of the Great Western Railway Company.

12. Mines and Minerals.

There is no mining in the usual sense of the word in Wiltshire. A little iron ore is dug at Westbury and smelted there. But as stone comes under the heading of minerals the great quarries of oolite freestone at Box and Corsham must be mentioned. The product of these is generally known as Bath stone, but it is of purely

Wiltshire origin and output. These quarries are of great
extent and some of them are worked in a peculiar manner.
A shaft is sunk, in some cases 100 feet deep, from which
galleries of considerable height are run in various directions. The surface of the earth is in consequence very
little disturbed in proportion to the great extent of the
industry, for in one of these underground quarries there
are over three miles of tramway. This is one of the
most durable, easily worked, and best known freestones
in England, and it is used all over the south for building
purposes. It was moreover the stone used for most of
the ashlar work in the churches of north Wilts in
mediaeval times, and indeed was worked long before this
by the Romans for their villas and walls.

The famous old quarries at Chilmark too are still
working, whence came the freestone used in the building
of Salisbury Cathedral.

13. History of Wiltshire.

The authentic history of Wiltshire begins dimly in
the latter half of the sixth century with the conquest of
its territory by the West Saxons under Cerdic. They
had already for a generation been advancing from the
south and south-west, and after many battles, some fought
against King Arthur, had established themselves over much
of modern Dorsetshire and even on the valley of the
Wily about Wilton. Cymric, the son of Cerdic, carried
on the slow but sure invasion. In 552 he crushed at

Old Sarum a large army of Britons, and four years later had swept on over Salisbury Plain and the Pewsey vale to the northern verge of the Marlborough down-land. He then completed the conquest of the local Britons in a fierce battle on Barbury hill where a still perfect en-

Wardour Castle, from the North

campment looks down over the vale of Swindon. The Saxons proceeded to another twenty years of fighting, which at the final battle of Deorham in Gloucestershire ended either in the subduing of the Britons or their expulsion to the Welsh border. But Wiltshire in 556

became permanently part of the kingdom of Wessex.
The British returned, it is true, and won a great battle
at Wanborough under Liddington camp near Swindon.
But it was of little use, and henceforth the struggles for
supremacy were between Saxon Mercia, which roughly
speaking was the central kingdom of England, and Saxon
Wessex. One of the greatest battles fought in these
wars was at Great Bedwyn, where the Mercians suffered
a severe defeat.

The Danes had begun to harry the English coast in
the eighth century. Throughout the ninth century and
particularly during its latter half they were continually
in the heart of the country and the energies of the men
of Wessex and their kings were almost wholly taken up
in endeavours to drive them out. For a long time they
were successful. But the Danes were better disciplined
and better armed than the Saxons, who moreover did not
always agree among themselves. The great Alfred now
came upon the scene, and after some rebuffs won his first
victory at Ashdown in Berkshire, near the Wiltshire
border. Soon afterwards, however, he and his brother
King Ethelred were defeated, as is thought, near Bedwyn,
Ethelred being slain. Alfred, now king, was almost immediately beaten by the Danes at Ethandun, perhaps Wilton,
after a prolonged fight. In 876 however he purchased
their retirement and a promise to abstain from attacking
Wessex again. King Alfred had now gained a long
experience of Danish fighting, having before these events
assisted the Mercians against the Norsemen, who had
conquered large parts of the north of England, and had

permanently settled there and accepted Christianity. But
the Wessex people had no desire for such neighbours,
excellent ones and good Englishmen as these northern
Danes in time became. Their experience of a Dane was
a cruel ravaging pagan. Alfred set himself to work on
more effective lines. He erected fortifications and built
a fleet which scoured the coasts and beat the Danes
on their own element, the sea, off Swanage. In 878,

Old Sarum

however, the Danes made the greatest of all their attacks
on Wessex, captured the royal town of Chippenham,
overpowered Alfred's forces, and compelled the Saxon king
to winter in the security of the Somersetshire marshes.
All seemed now lost. But Alfred emerged in the spring,
gathered an army together, and fought the memorable
and decisive battle of Ethandun, thought to be Edington
near Westbury. The Danes were forced by this to
evacuate Chippenham and conclude the Treaty of Wed-

more, in which they agreed to pay homage to Alfred, evacuate Wessex, and in the person of their leader Guthrum and his chief to accept Christian baptism. All these conditions were carried out, Alfred continued his defensive precautions and Wessex had peace. But in the next century the feebleness of some Saxon kings laid Wessex open to more Danish wars till the Norman Conquest in 1066 put an end to their invasion and brought into utter subjection both the Saxon and Danish inhabitants of this island.

After the battle of Hastings Wiltshire, together with the greater part of England, came at once under the Norman yoke. The great Saxon landowners of the county, like others, were deprived of their estates. These were bestowed on their conquerors, who by degrees built castles upon them and introduced the feudal system. The lot of the common people, whether serfs, partly serfs, or freemen, was not perhaps very greatly altered, but the Thanes of Wiltshire became mere vassals under the great Norman lords and were regarded as inferiors. William the Conqueror in 1070 held a great muster beneath Old Sarum of the army to which he owed his conquest, dealing out rewards to its chiefs ; while again, just before his death, he here assembled all the leading men of the whole kingdom, who then renewed their oath of fealty to him for their lands. It was here, too, that William gave orders for the preparation of the famous Domesday book, a record of this same distribution of lands, which at this day tells us to a great extent both their old and their new owners. During the wars of Stephen and Matilda

Wiltshire more than most counties was the scene of fighting and ravaging for fifteen years, Devizes, Marlborough, Malmesbury, and Trowbridge and their castles, together with Wilton and Salisbury, playing a leading part.

Though it continued to be a very important county, constantly the abode of kings and queens at their royal castles such as Devizes and Marlborough, and the seat of many wealthy religious establishments and of great nobles, with whom it was popular for its hunting, which, next to war, was then the most important pursuit among the great, Wiltshire remained generally quiet. The Wars of the Roses, though destroying for ever some of its great families, did not actually disturb the county so much as many other parts of England, nor did any of the peasant insurrections of the middle ages seriously affect its peace. By the influx of Flemings, as elsewhere mentioned, it became a principal seat of English cloth manufacture, and after the dissolution of the monasteries, such as Wilton, Amesbury, Lacock, and Malmesbury, the wide church lands passed into the hands of families, some old, some new, who were either favourites or offered sufficient money for them at the time.

Once again, in the struggle between King Charles I and Parliament, Wiltshire became an active seat of war. Though the west generally sympathised with the King's party while the east of England mainly took the other side, Parliament had a strong following in Wiltshire, particularly in the towns. Marlborough especially was conspicuous as a Roundhead centre. As Wiltshire stood

Salisbury Cathedral, Choir looking West

on the road not merely between London and the west but also near Oxford, the King's headquarters for so long, it was natural that there should be a great deal of fighting in the county. The two battles of Newbury took place just outside its borders on one hand while the frequent struggles for Bristol were equally near to it on the western border. For a long time things went very badly in Wiltshire, as elsewhere, for the Parliament. Cromwell and his " new model," his disciplined regiments largely from the east of England, had not yet come to the front. Neither side had much discipline, but the Royalists having most of the gentry and their immediate followers were more used to arms as well as better equipped and better led. Sir E. Hungerford and Sir E. Baynton however, heads of two ancient and powerful Wiltshire families, led the local Parliamentarians, but not very successfully, as they fell out with one another. In 1642, the first year of the war, Marlborough, after severe fighting, was taken by the Royalist troops, who caused much loss and suffering there. Malmesbury, on the other hand, held for the King, was taken in 1643 by Sir William Waller after some fighting, but frequently changed hands. The siege by Hungerford of Wardour (near Tisbury), the only ruined castle now standing in Wiltshire, is famous, as it was defended by Lady Arundel herself in the absence of her husband the owner. With 50 men, only half of whom were soldiers, she withstood the attack of 1300 Parliamentarians with artillery for five days, the women loading the muskets. It was then held for a year by Ludlow, a native of south Wilts, and later a prominent

Parliamentary leader, to be recaptured by the son of the brave lady who had defended it, her husband in the meantime having been killed at the battle of Lansdowne. Devizes saw the most important battle fought in Wiltshire during the war. For in 1643 the Royalists under Lord Hertford and Prince Maurice sought safety and

The Old Seymour House, now Marlborough College

defence there after their defeat at Lansdowne. Waller followed them and invested the town, but news of its plight being carried by horsemen to Oxford, Lord Wilmot with 1500 horse arrived on Roundway down, just above Devizes. Upon this Waller led his army up to meet him and was most ignominiously beaten and put to flight, leaving all his guns, baggage, and 2000 killed and prisoners behind him. Salisbury was alternately occupied by both sides during the war, and in 1645 there was a smart fight in its streets between Ludlow, holding it for the Parliament, and Marmaduke Langdale, who was endeavouring to drive him out. One of the last engagements in Wiltshire was the storming of Devizes castle by Cromwell in the same year. After the battle of Worcester, when Charles II was making his way to the coast, he lay concealed for some days in Heale House between Salisbury and Stonehenge. Salisbury also witnessed the abortive rising of Penruddock and other Wiltshire gentlemen against the Parliament. They seized the judges then sitting at assize and proclaimed Charles II, but met with no support and were soon caught and executed.

Wiltshire also saw the march of William of Orange and his army through its borders, and more than that. For James II had assembled his army to oppose the Prince at Salisbury, making his quarters at the bishop's palace. But the king retired and his son-in-law William occupied the palace for several days, Salisbury then becoming a centre of negotiation and intrigue among the leaders of the kingdom and their friends. It was while at Littlecote in north-east Wilts that William assembled the notables of both parties and received the news of his father-in-law King James' flight from London. Nothing further of national importance took place within the borders of Wiltshire. The riots occasioned in England by the introduction and improvement of machinery both in factory and field, at a time when labour was low and plentiful early in the nineteenth century, were serious in the county because the cloth weavers were still an important element in north-west Wilts, and the field labourers were more numerous and worse paid than in any county except Dorsetshire, which stood on the same low level.

14. Antiquities—Prehistoric, Roman, Saxon.

In no county probably are there so many prehistoric mounds or barrows as in Wiltshire and, owing to the smooth open turf down on which they mostly occur, in none certainly do they over so great an area form so conspicuous a feature in the landscape. These ancient graves cluster, as is natural, most thickly about the great temple of Stonehenge on Salisbury Plain and Avebury on the Marlborough Downs, though existing almost everywhere, while numbers have been levelled by the plough. The barrows are of two distinct varieties, the round and the long. The latter are far the larger and belong to an earlier age than the more numerous round barrows. They vary from 100 to nearly 400 feet in length, 30 to 50 feet in breadth, and nowhere have a height in their present condition of over 12 feet. There are some sixty long barrows in the county and by rough computation about 2000 of the others still existing. The former, which generally lie east and west, have the broadest and highest end towards the east. Many have been carefully excavated and in the districts where sarsens exist found

Palaeolithic implement
(From Kent's Cavern, Torquay)

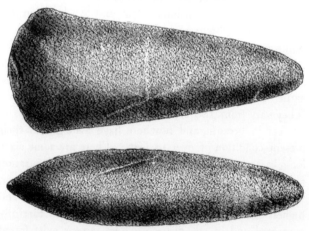

Neolithic Celt of Greenstone
(From Bridlington, Yorks.)

Stonehenge

to contain chambers made of these stones, either at the east end or at the sides, with sometimes passages of sarsens leading to them. In these chambers entire skeletons have frequently been found, usually in a sitting position, together with many flint implements. The skulls found in these long barrows are longer than those of any present European race. These men of the neolithic or late stone

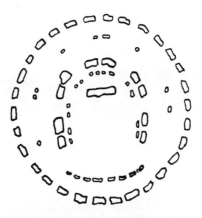

STONEHENGE RESTORED

age, the "long-headed men" of dark hair and complexion, were conquered eventually by fresh invaders, a race of larger physique, light hair, and round heads, who knew the use of bronze, and buried their dead in round barrows. Where there are no sarsens, as about Stonehenge, the long barrows are nothing but chalk and earth, with a deep ditch dug along either side, but not carried round the ends. In some of them numbers of skeletons thrown in promiscuously near the higher and broader end have been found. The skulls of many have been obviously fractured before death, and though they might have been the victims of a battle many hold the theory that they were the wives and slaves of some great chief buried with him.

Martinsell, showing Remains of old British Village

In the round barrow the skeletons have often been found in a crouching attitude like those of the long-headed neolithic men in the chambered long barrow, but custom changed and cremation became the fashion with these people of the bronze age, their ashes being frequently found in rough earthenware urns, though the burial of

bodies was never entirely given up. Both forms of burial
are sometimes met with in the same barrow. Earthen
cups, too, of various shapes are found in the round
barrows, and from some have been taken gold ornaments,
buttons, bosses, and beautifully made bronze dagger
blades, with awls or prickers, and other relics. In the
long barrows only flint and bone implements, arrow-heads,
scrapers, hammers, and such like have been discovered.
There are many fine collections in Wiltshire, public and
private, of these prehistoric relics, notably in Salisbury,
Devizes, and Marlborough. It is now thought that the
use of bronze may have been introduced into Wiltshire
about 1800 B.C. and continued in use until that of iron
came gradually in, about 300 B.C.

But the monster mound of Wiltshire and indeed of
Western Europe is Silbury hill, close to Avebury. This
gigantic tumulus is 130 ft. high, and its base covers no
less than 5 acres of ground. Two or three serious exca-
vations have failed to yield any clue to its secret, and it
is, and perhaps will always remain, an object of speculation to the antiquarian as to its origin and purpose, and
as a prodigious monument of human industry. One of
the next largest in England, and raised perhaps for the
same mysterious and unknown purpose, is in the college
grounds at Marlborough, and was the seat of the Norman
castle which once stood there.

Wiltshire, however, contains the three great prehistoric monuments of Britain. Silbury has been mentioned. Stonehenge and Avebury, the former particularly,
are better known. Stonehenge, some eight miles north

of Salisbury, happily in a still comparatively wild and open part of the Plain, has, unlike Avebury, had a wonderful measure of escape from the destroyer's hand. Everyone is familiar from pictures at least with its general appearance. Its formation was an outer circle of roughly-squared large stones about 15 feet high, along whose tops were laid and fastened with rude mortices other great

The " Devil's Den "; a Dolmen near Marlborough

stones known as imposts. Next was an inner circle of smaller stones about 6 ft. high. Within these circles were two ellipses (horse-shoe shaped), the outer one composed of detached trilithons (two huge uprights 25 ft. high with imposts), and within it one of smaller stones, with—innermost of all—a large, not quite central slab of sandstone. There are also single monoliths at the approach and sides. The whole temple is surrounded by a circular ditch. About half of the great outer stones (16) still stand, a few of them still bearing their imposts. Two of the five huge inner trilithons with imposts are perfect, with the others either as upstanding monoliths or lying prone. Of the smaller stones forming the inner circle and inner ellipse,

Avebury Stones

being more easily carried away, a much smaller proportion remain. All kinds of legends are current, some connected with Merlin, regarding the bringing and raising of these great stone pillars. But there is nothing really mysterious in such a performance. The outer and longer stones and inner trilithons are sarsens and at the furthest came from the Marlborough Downs, no impossible feat of transport for great bodies of men with ropes and levers to whom time was little object. The smaller stones are of different igneous rocks, probably from South Wales or North Devon, but not large enough to present even thus any serious difficulty in transport.

The fabric is almost certainly connected with sun worship, the rising sun on midsummer day striking a line on certain stones, obviously placed there for the purpose. The movement of the earth in respect to the sun since these astronomical marks were set there, enables some modern astronomers to think that they can fix the date of their erection, or in other words the date of Stonehenge. This works out at between 1400 and 1600 B.C., a period which moreover agrees with the stone implements evidently used in the construction which were found only a few years ago beneath one of the fallen monoliths which was being raised to its original position. Who were the builders, however, will never be known.

Avebury, between Marlborough and Calne, is by some considered older than Stonehenge. The temple site is of huge size—a great circle over three-quarters of a mile in circumference and 28 acres in area, surrounded by an enormous earthwork about 70 feet in height, with a fosse on its *inner* side. All this area is now occupied by the village of Avebury, which dates back to Saxon times, and was practically constructed out of the broken fragments of the great monoliths forming the temple. The arrangement of the temple differed considerably from that of Stonehenge. All round the circumference of the enclosed area, just within the fosse, were placed upright unhewn sarsen stones, about 100 in number and over 12 ft. high, of which a few stand, while some are prostrate, others are buried, and the remaining five-sixths destroyed.

The Font, Avebury Church

Within were two small circles, side by side, of about 30 stones, equally large or larger, of which scarcely a dozen remain either upright, prone, or buried. From this great temple, to which Silbury is immediately adjacent, an

avenue consisting of upright sarsens ran south to West
Kennet village, and probably further to a smaller circle
on Overton hill, but very few of these now remain. A
second avenue is thought by some to have run to Beck-
hampton. Avebury had probably no rival in England,
and possessed an importance impossible to gauge at a
date we cannot estimate. On the smooth down around
it, as at Stonehenge, the barrows lie with exceptional
thickness, suggesting that its neighbourhood gave a
peculiarly sacred character to the ground. Aubrey, the
Wiltshire historian and antiquary, first drew modern
attention to Avebury in 1648, and left a plan of it which
shows about seventy stones standing out of the two
to three hundred which composed the original temple.
Eighty years later Dr Stukeley, who held serpent-worship
theories about Avebury, saw the stones being rapidly
demolished for local building purposes. So not only
splendid abbeys and churches were destroyed by our
thoughtless or bigoted forefathers in the sixteenth and
seventeenth centuries, but these solemn memorials of
mysterious ages were also treated by all classes alike as
of no account. Of cromlechs or dolmens, i.e. massive
stones supporting a roofing-stone and known to be
sepulchres, the Devil's den, near Marlborough, is a comparatively perfect example.

There are two conspicuous lines of dyke, or deep dry
ditches of unknown origin, traversing Wiltshire, besides
lesser ones. These are Bokerly dyke in the south of the
county, very noticeable near Martin, and the much more
famous work—the Wansdyke (Wodens-dyke). This

last enters the county near Great Bedwyn and runs over
the Marlborough Downs, where for many miles its high
rampart and deep ditch are conspicuous from afar, cleav-
ing its way up hill and down dale like a huge furrow.
Leaving the high down between Roundway and Calne
it crosses the lower country to the neighbourhood of
Bath, only here and there distinguishable.   Its purpose is
a matter only of conjecture, but like Offa's dyke it is almost
certainly a boundary.   A now discredited theory holds
it to be the limit of the Belgae, the last invading tribe
before the Saxon influx.   But most incline to the opinion,
formed by General Pitt-Rivers from coins and relics
found in them, that both this and Bokerly dyke are of
Roman, or Romano-British, or even Saxon construction.
Camps, occupying almost always the summits of the
down escarpment overlooking the vales, are extremely
numerous in Wiltshire, both circular, irregular, and
roughly rectangular, belonging no doubt to widely sun-
dered ages.   They invite much thoughtful but futile
speculation and much controversy.   It will be enough to
say here that they were not only the lofty strongholds of
tribal peoples against their enemies, but places of refuge
in all probability for their live stock at night against the
numerous beasts which haunted the tangled forests below.
They were doubtless used in later times, and often
adapted to the current needs both of Roman, Dane,
and Saxon in their wars.   The most noted of them are
Barbury and Liddington looking north from the Marlborough Downs; Bratton, Ogbury, Sidbury, Casterly,
Yarnbury, Battlesbury, and Scratchbury looking north

and west from Salisbury Plain. Ancient trackways leading from the downs to the low country, beaten by the hoofs of cattle for centuries into deep hollow trenches, are numerous, while everywhere on the slope of the downs are "lynchetts," i.e. terraces made by cultivation in time unknown. Then there are other trackways not merely for animals going down to water or pasture, but through-routes of traffic, used by those whom we for convenience call the ancient Britons. These always kept to the heights, while the Romans usually avoided them. The turf ridge-way running along the north escarpment of the Marlborough Downs is a well-known example. Of Roman stations there were not many of importance in Wiltshire. Sorbiodunum (Old Sarum), Cunetio, a mile to the east of Marlborough, and Verlucio (Heddington) near Calne, were perhaps the most conspicuous. Of Roman roads many traversed the county. One went from Bath to Heddington and Marlborough, while several diverged from Old Sarum.

Though the Saxon era begins the period of authentic history the solid remains they have left us are comparatively scarce and come under the category rather of Architecture. No doubt the traces of the Saxon pick and shovel in hill-top camps in the alteration of old ones or otherwise are before our eyes, as are those of the Romans, but we cannot identify them. The enormous trenches that surround the great hill of Old Sarum are attributed to a Saxon king. But the Saxon built mainly of wood. His roads probably followed the lines of the British and Roman roads, and his earthworks we cannot identify or separate from those of others. One exception, however, should be mentioned, and that is the Moot hill at Downton, which seems most probably to have been raised by the Saxons as a meeting place for the shire mote or local parliament.

Saxon Church, Bradford-on-Avon

15. Architecture—(*a*) Ecclesiastical.

A preliminary word on the various styles of English architecture is necessary before we consider the churches and other important buildings of our county.

Pre-Norman or, as it is usually, though with no great certainty termed, Saxon building in England, was the work of early craftsmen with an imperfect knowledge of stone construction, who commonly used rough rubble walls, no buttresses, small semi-circular or triangular arches, and square towers with what is termed "long-and-short work" at the quoins or corners. It survives almost solely in portions of small churches.

Bradford-on-Avon Church : interior

The Norman conquest started a widespread building of massive churches and castles in the continental style called Romanesque, which in England has got the name of "Norman." They had walls of great thickness, semi-circular vaults, round-headed doors and windows, and lofty square towers.

Norman Door; St Mary's, Marlborough

From 1150 to 1200 the building became lighter, the arches pointed, and there was perfected the science of vaulting, by which the weight is brought upon piers and

buttresses.   This method of building, the "Gothic,"
originated from the endeavour to cover the widest and
loftiest areas with the greatest economy of stone.   The
first English Gothic, called "Early English," from about
1180 to 1250, is characterised by slender piers (commonly
of marble), lofty pointed vaults, and long, narrow, lancet-
headed windows.   After 1250 the windows became
broader, divided up, and ornamented by patterns of
tracery, while in the vault the ribs were multiplied.   The
greatest elegance of English Gothic was reached from
1260 to 1290, at which date English sculpture was at
its highest, and art in painting, coloured glass making,
and general craftsmanship at its zenith.

After 1300 the structure of stone buildings began to
be overlaid with ornament, the window tracery and vault
ribs were of intricate patterns, the pinnacles and spires
loaded with crocket and ornament.   This later style is
known as "Decorated," and came to an end with the
Black Death, which stopped all building for a time.

With the changed conditions of life the type of
building changed.   With curious uniformity and quick-
ness the style called "Perpendicular"—which is unknown
abroad—developed after 1360 in all parts of England and
lasted with scarcely any change up to 1520.   As its name
implies, it is characterised by the perpendicular arrangement of the tracery and panels on walls and in windows,
and it is also distinguished by the flattened arches and the
square arrangement of the mouldings over them, by the
elaborate vault-traceries (especially fan-vaulting), and by
the use of flat roofs and towers without spires.

The mediaeval styles in England ended with the dissolution of the monasteries (1530—1540), for the Reformation checked the building of churches. There succeeded the building of manor-houses, in which the style called "Tudor" arose—distinguished by flat-headed windows, level ceilings, and panelled rooms. The ornaments of classic style were introduced under the influences of Renaissance sculpture and distinguish the "Jacobean" style, so called after James I. About this time the professional architect arose. Hitherto, building had been entirely in the hands of the builder and the craftsman.

Reverting again to the term *Saxon* a great many Wiltshire churches contain undoubted fragments of that period. Manningford Bruce, though Norman in style with an apsidal chancel, is thought to ante-date the Conquest. But at Bradford-on-Avon there exists the finest and most perfect specimen in England of an unaltered Saxon church of the tenth century at the latest. It is very small, gloomy, and rude, consisting of a nave, chancel and porch.

The noble Abbey church of Malmesbury, still in great part perfect and in use, has the finest Norman work of any, though mingled with transition. The two churches of St John and St Mary Devizes have also much beautiful Norman work, while among the smaller churches, Preshute and Manningford Bruce are conspicuous examples of the Norman style.

Salisbury Cathedral is the finest specimen of pure Early English existing anywhere and has the tallest spire in the kingdom. Peculiar interest attaches to this

cathedral, as before its erection a predecessor had stood
for two centuries on the top of the hill of Old Sarum,
and when the town was moved down to the present site
the material of the old cathedral was brought down also
and used in the later building.   As types of the Early

Interior, Avebury Church

English style Bishop's Cannings and Potterne are among
the best in Wiltshire, while, though more mixed with
other periods than these, Cricklade, Amesbury, Downton,
St Martin's in Salisbury, and Great Bedwyn, with several
others, are also good examples.   There is not a great deal
of the fine tracery of the Decorated period in Wiltshire,

but the tower and transepts of Lacock church afford one instance at least to the contrary.

The Perpendicular style, as already mentioned, is very prominent in the county, which grew rich rapidly by wool-growing about that time and had money to spend on its churches. The exterior of a church, either

Amesbury Church

Perpendicular in origin or by restoration, will be the most familiar of all styles to Wiltshire people and is obvious at a distance. It is generally battlemented throughout, with pinnacles not only on the tower but often elsewhere, these being crocketted, i.e. decorated with imitation sprouts or knots. In the chalk districts of the county flint was used and, owing to the lack of good stone, in much of this district the churches are inferior in size and dignity to those of the lower country. The Perpendicular style exists in the whole or portions of Edington, St Mary's Devizes, Bradford, Trowbridge, St Sampson's Cricklade, St Peter's Marlborough, and Mere. The first-mentioned is one of the most remark-

Edington Church

able examples in England of transition from Decorated to Perpendicular.

The principal monastic houses in Wiltshire were at Wilton, Malmesbury, Amesbury, Stanley, Lacock, and Bradenstoke. They were swept away by Henry VIII. Wilton was founded in the ninth century by King Alfred for Benedictine nuns, and was very rich when its property was granted by King Henry to the ancestor of the Herbert family. The former owners of Amesbury were a nunnery founded in the tenth century and afterwards much associated with the family of King Edward the First. The buildings of both of these have practically vanished.

Bradenstoke, or Clack, was founded in the twelfth century for Black Canons, and there are still considerable remains of the abbey. Malmesbury was founded in the eighth century, but very little of its conventual buildings remain around the magnificent abbey church already spoken of.

Stanley Abbey was founded by the Empress Maud for Cistercians in the twelfth century, and there is nothing now left of it but its grass-grown foundations. Lacock was founded as an Augustine nunnery by Ella Countess of Salisbury, who became its first abbess. There are considerable remains of the conventual buildings, with their beautiful cloisters in good preservation, attached to the later residence of the Talbot family, forming a noble whole.

There are two remarkable specimens of modern church architecture in Wiltshire, the Lombardic church built by the Herberts at Wilton, and the College chapel at Marlborough, the latter regarded as perhaps the most beautiful of the kind in England. Two churches in the county present the curious spectacle of two steeples, one in the centre and one at the west end, namely, Purton and Wanborough. Finally, it should be stated that Ramsbury was the Diocesan See of Wiltshire for a considerable period just before the Conquest, Salisbury assuming that position in 1075. Marlborough was a suffragan bishopric in the Tudor period, and the title was revived in recent years, but only for application to one of the suffragan bishops of London, and it has now been discontinued.

16. Architecture—(*b*) Military. Castles.

There were scarcely any castles in England before the Norman Conquest, but the Normans, to secure their conquest, erected between one and two thousand. Some were built as royal fortresses by the Norman kings for the defence of the country and of their own interests against the ambition of their subjects. Others were built by the great Norman landowners to secure their individual territories against their neighbours and overawe the conquered Saxon people. Old Sarum, Marlborough, and Ludgershall were Royal castles at the disposal of the Crown or held by constables; Devizes, on the other hand, together with other equally magnificent ones in neighbouring counties, was built by Roger, Bishop of Salisbury. All of these were set on the top of great mounds, some natural and some artificial. Great architectural skill as time went on was expended in planning and building them. The earliest often had square towers, a style which was subsequently abandoned for round ones. The plan of these castles consisted of two separate areas of defence, one within the other, each defended by walls and towers and the whole surrounded by a moat. Within the inner defence or "inner bailey" were the most important buildings, and chief of all the strong "keep," with a well in it, as a last resort for a hard-pressed garrison. The barons' castles, always garrisoned by armed retainers, were a terror to the country, where with the exception of the monasteries there was scarcely anything larger than a wooden manor house of four or five rooms. The condition on which the vassals held their lands included specified services in defence of the castle. After the Stephen and Matilda wars the Wiltshire castles did little more than act as a security against any rising of the people and the aggressions of other barons. Neither in the barons' wars of Henry III nor in the Wars of the Roses was there very much fighting in the county. When the Civil War of the seventeenth century broke out and forced so many old castles, by that time partly remodelled as country houses, to defend themselves, Wardour, already spoken of, was almost the only true fortress house in Wiltshire, and its ruins are now, save a fragment at Ludgershall, the solitary example in the whole county of a feudal castle. It is curious that a shire so rich in the memorials of the past of nearly all periods should have lost practically all its castles, but in looking at its history there is some reason for it. Wiltshire is neither on the sea where foreign or piratical attacks might be expected, nor is it near the borders of Wales, where the Welsh were aggressive and the border barons exceptionally powerful and quarrelsome among themselves. After the middle of the twelfth century Wiltshire, compared with many counties, seems to have been quiet till the Civil Wars. So its castles dropped out of use and into decay, and their stones were often used for the building of houses, barns, and walls.

17. Architecture—(c) Domestic. Famous Seats. Manor Houses. Cottages.

Wiltshire is very rich in domestic architecture. As we have seen, after the Wars of the Roses, during which the powerful baronage of England half destroyed itself, and the Tudor kings to strengthen their own power did all they could to prevent its recovery, the need for strongly fortified castles ceased. A new kind of aristocracy, raised up from the ordinary country gentlemen and the wealthier merchants, was encouraged by Henry VII, Henry VIII, and Queen Elizabeth as a counterpoise to the old turbulent and proud barons who had been always a menace to the Crown ; a class greatly strengthened at the Reformation by the distribution among them by gift or easy sale of the vast church lands then confiscated.

The castles were now by degrees remodelled as country houses or entirely replaced by great fabrics built in what is broadly known as the Elizabethan style. Bricks, hitherto little made, as well as stone and wood were all used, in a degree more or less dependent on the material which the district most readily produced, often helped out by the stones of the old fortresses where such had existed. The usual style was a quadrangle surrounded by buildings, for precautions against possible trouble were not wholly abandoned. Instead, however, of gloomy and in many ways uncomfortable places of residence, intended to harbour large followings of armed men, covering many acres and proof against artillery, there sprang up in their places family residences of much beauty, letting in light and air to the greater rooms through splendid windows.

South Wraxall Manor House

Large retinues were still often maintained but they were mainly for show and pomp. Besides the great class of the nobility, there now arose a minor aristocracy that may be described as ordinary country gentry. These had always more or less existed in the feudal period, but rather as followers and semi-dependents of the great baronial houses. They had lived very plainly in small rude manor houses, generally of wood ; not greatly differing in degree from the class just below them of considerable farmers and rural traders, between whom and themselves it would not have been always possible to draw the line. All these middling people had been a great source of strength to England and were far more independent than any similar class in any other country. Theoretically

Old Tythe Barn, Bradford-on-Avon

they represented the old Saxon gentry, as the feudal aristocracy with much more accuracy had represented the Norman conquerors. This question however is very complicated, but as a broad truth it may be accepted that the independent country-gentleman class, as the word is now understood, came into being in the sixteenth century. They rebuilt or embellished their houses by degrees in a modest way just as the great families did theirs in a more splendid fashion, while a large proportion of the latter had themselves, as we have seen, risen out of the lesser gentry. Wiltshire was rich in both kinds and consequently in good houses. Scarcely any county possesses more, very few so many, fine specimens of early domestic architecture.

One of the oldest known domestic buildings in the county is the huge fourteenth century barn and farm house at Bradford. Norrington, near Ebbesborne, is a fine example of a very early fifteenth century manor house, originally owned by the ancient but extinct family of Gawen, and so also is Tollard Regis and a notable timbered house in Potterne. South Wraxall manor house, the cradle of the well-known Wiltshire family of Long, is not only fifteenth century but is an ample and fine specimen of a country house of that period, though remodelled a little later. Great Chalfield has beautiful portions of a manor house of mid-fifteenth century date. There are also several houses in Salisbury cathedral close and city of that century, while of great houses Littlecote, once the property of "Wild Dayrell" and afterwards till the present day of the Pophams, is one of the most perfect Elizabethan mansions in England. Kingston House at Bradford, late Elizabethan, was selected as a model of that period for representation at the last Paris exhibition. Longleat, the seat from its first erection of the Thynnes, was built in the late sixteenth century by an Italian architect and was then one of the noblest houses in England. Charlton, built by Sir J. Knyvet, is another fine example of a little later period. Corsham the seat of Lord Methuen, Longford that of Lord Radnor, and Wilton that of the Pembroke branch of the Herberts, all of this period, have much more than local fame. As very perfect specimens of smaller Elizabethan mansions Lake House on the Avon north of Salisbury, and the manor

Littlecote

house at Stockton, are not perhaps excelled in the county. In great and stately houses of later date, chiefly eighteenth century, Wiltshire is equally rich, Stourhead and the later Wardour Castle being fine examples. Of the more imposing Queen Anne mansions the old seat of the Seymours, now the main building of Marlborough College, is the most interesting. Another fine specimen in the same style and also of red brick is Ramsbury manor house of the Burdett family. This Queen Anne style— a rectangular building with long regular windows under a single high-pitched roof without gables but with projecting eaves—is frequent in Wiltshire, as elsewhere, but more prevalent on a small scale in the former residences of prosperous citizens in country towns. The above are but a few examples of the many old houses that abound in the county.

What is often called "farmhouse" architecture nearly always comes under the category treated of above. That is to say most of the beautiful old farmhouses of the Tudor, or the Jacobean style which followed it, as well as some of the later Queen Anne houses, were originally manor houses and the residences of the lesser country squires who were bought out by their wealthier neighbours in the eighteenth and early nineteenth centuries. For this was a period in which great estates became greater, and the lesser gentry sometimes burdened with large families and tempted by high prices disappeared in great numbers all over England. In the extreme north and north-west of the county, where Cotswold and Bath stone was convenient, these middling-sized manor houses are particularly handsome. In the chalk countries flint and sarsen stone were often used, though stone from Corsham and Chilmark and other noted quarries, as well as brick, was also constantly employed.

Graceful from an artistic and architectural point of view as are all these styles, when exhibited in great or small country mansions or in genuine farmhouses, the humbler cottage architecture of a country is almost more important as a feature of its general life and scenery. Wiltshire in this respect may challenge comparison with most of the southern half of England, which is beyond all comparison more picturesque and interesting than the northern half, drawing a rough line across the map from Cheshire to Norfolk. The old cottages of England though often not so healthy and roomy as those built nowadays, which in Wiltshire are generally of red brick and always ugly, are nevertheless most interesting as being part of the old life of their counties and reflecting their characteristics in the days before the easy transportation of materials began to destroy these marked differences in districts. In the chalk regions of the county particularly, almost entire villages with thatched roofs, covering walls sometimes of white-washed mud, sometimes of timber and wattle, or again of flint and sarsen, are very common. Along the valley of the Avon through Salisbury Plain, or along those again of the Wily or the Ebele in south-west Wilts, these may be seen in perhaps the greatest perfection from the picturesque point of view. But, as a matter of fact, Wiltshire being almost entirely an agricultural county, and save in a few places having no industrial or outside demand for cottages, these old-fashioned villages are the rule rather than the exception throughout the county, upland and lowland. Shaded often by immemorial elms, and generally containing one or two large farmsteads with their well filled stackyards, they present a peculiarly snug appearance, more especially when contrasted with some wild stretch of down rising high on either side of them. When we get to the northern fringes of Wiltshire we meet an entirely different style of architecture both in cottage and farmhouse, namely that of the yellowish grey stone which is more generally identified with the Cotswold districts of the adjoining counties of south-west Oxfordshire and Gloucestershire.

The Hall, Bradford-on-Avon

This is most striking and has even more character. For in mid and south Wilts the flint cottage, often built in a fancy pattern of chessboard-like squares, is the only style that may be called a really local one, since the others are common to much of the southern half of England. But the villages of Cotswold and Bath stone which north Wilts shares, though in a minor degree, with its neighbours have a character and reputation of their own among all lovers of British domestic architecture.

These differences show themselves in the towns as well. Bradford for instance is an almost purely stone town, though among the stone buildings are many old half-timbered houses of extreme antiquity. It is quite different

The Moot House, Downton

from any other considerable town in Wiltshire in appearance, and makes comparison with others impossible. Salisbury, for the number of ancient houses in its streets, further embellished by its matchless cathedral close with its own architectural treasures and by the many clear chalk streams purling through and round the city, also stands somewhat alone. Of the rest Marlborough with its long wide sloping street, its many old buildings, its penthouse and delightful surroundings is perhaps the most admired by strangers, and this too in spite of the fact that three-fourths of the town was destroyed by a disastrous fire in 1653 and rebuilt. Malmesbury, terminating in its noble abbey, has a distinctly old-world look. In all the other Wiltshire towns—unless it be Highworth perched on a lofty ridge looking down into Berkshire and very little altered—the modern builder and modern needs have to a great extent banished the old-time atmosphere, though this remark does not apply to very small towns, in reality large villages, such as Heytesbury, Great Bedwyn, Aldbourne, or Amesbury, nor again to Castle Combe on the Cotswold slopes near Bath, which is sometimes held to be the most beautiful village in Wiltshire. But Lacock for its profusion of old half-timbered houses and remarkable atmosphere of antiquity is unequalled from a purely architectural point of view among the larger villages of the county.

18. Communications—Past and Present.

The old Roman roads are still visible in many parts of Wiltshire, and in others underlie the modern roads. They were elaborately constructed with many layers of stone, broken up or otherwise, and sometimes even mortared. Except that these roads were much narrower than ours almost as much trouble was taken with some of them as is expended nowadays on the streets of a big town.

Of those in Wiltshire the Via Julia from Bath by
Verlucio (Heddington) to Cunetio (near Marlborough)
and thence to Silchester (Berkshire) is well known.
Another ran from Sorbiodunum (Old Sarum) N.E. to
Silchester, and a third from the same place to Winchester.
There is also a road from Salisbury to Blandford (Via

Downton Village

Iceniana), while Cirencester and Marlborough were connected by another.

It seems curious that from Roman times till about a
hundred years ago the roads of England were of a kind
such as the Romans would have regarded, and we should
now regard, as barbarous. The stages of slight improvement which marked the eighteenth century are hardly

worth consideration. A return to something like the
Roman method, that is making a stone bed to a road and
using broken stone, was not achieved till Macadam introduced it and till fast stage coaches inaugurated a new
era of travelling. If a modern Wiltshireman could see
even his famous Bath road in winter weather in the time
of George II, or yet later, he would be amazed. In those
days roads were merely dirt tracks, levelled after the winter
traffic with the plough, drained in a sort of way with
water furrows, sometimes patched with unbroken stones,
logs or brush, and then left to the mercy of the traffic and
the weather. In summer they got beaten fairly hard,
though into a very rough surface. In winter or continuous wet weather they were cloven into ruts deep
enough to upset vehicles, and were often a foot or two
deep in mud in which carriages frequently remained im-
moveably imbedded. The degree of badness depended on
the nature of the soil, whether sandy, limestone, or clay.
Three miles an hour was the usual rate of progress in
winter for a great person with a coach and no stint of
horses. In summer it was about five, while occasionally
the main roads became impassable during several consecu-
tive days for wheeled traffic. People rode on horseback
for preference on this account, and pack-horses laden
heavily and driven in long strings were in many parts of
the country a regular means of transport. Goods traffic
was also carried on by immense waggons drawn by four
or six horses. People who had no horses or carriages and
could not afford the post-chaises that were for hire,
travelled in these and in stage waggons or coaches whose

advertised speed was three miles an hour. But, as may be imagined, they did not leave home more than was necessary. It will be understood, too, how much more stay-at-home every one was in those days, and how truly each man's county or district, if an inland one, was his world. If there was a lean harvest in Wiltshire, for instance, it

Cerney Wick Lock and Round House

meant hard living for the poor and high prices, though the eastern counties might have a glut of corn. Canals came in towards the end of the eighteenth century, while roads, though considerably improved, had still to be made in either the modern or the Roman sense. It is needless to state what a boon water-carriage promised to be to an inland county. Great excitement and great speculation accompanied the introduction of canals. But the proper construction of roads, followed so quickly by the introduction of railroads, and all coming so soon afterwards, neutralised to a great extent the effect that canals would have had and the relief they promised to a people who had such difficulties in getting their produce to market. The Wiltshire canals, however, are still very useful in a subsidiary way. One of them, uniting the Thames and Severn, and opened in 1789, touches Wiltshire near Cricklade. The Wilts and Berks canal from the Thames goes through Swindon, and passing near Wootton Bassett and Chippenham, joins the Kennet and Avon canal south of Melksham. The latter water-way runs hence by Devizes and through the Pewsey Vale into Berkshire at Hungerford. The North Wilts canal unites the two first mentioned, passing from Cricklade to Swindon.

The main lines of the two companies which compete for the west country traffic, the Great Western and the South Western railroads, pass through north and south Wilts respectively. The former sends two main routes from Reading through north Wilts, the old one by Swindon and Chippenham, and a later line, somewhat shorter, through Hungerford and Savernake to Westbury. The Swindon and Andover line, via Marlborough, now connects north and south Wilts on the east, just as the line from Chippenham, Trowbridge, and Westbury down the Wily valley to Salisbury connects them on the western side. A line from Swindon to Cheltenham opens direct traffic into Gloucestershire and Worcestershire, while

Salisbury is connected by separate lines with Portsmouth to the south-east, and Wimborne and Bournemouth on the south-west.

The Great Western planted their works at New Swindon, and upon a site that within easy memory, as related elsewhere in this book, was green fields, have created a town of some 50,000 inhabitants, the only really large town in Wiltshire.

Of road arteries, now again so important, the Bath road, entering the county at Hungerford and passing through Marlborough, is probably the best known and most travelled. Another important line of road from north to south runs from Cricklade to Swindon, then climbing the Marlborough Downs and passing through the plateau of Savernake forest, descends into the Pewsey Vale and skirts the eastern edge of Salisbury Plain to Ludgershall, once a remote village and now a busy entrepôt of the permanent camps near by. There this fine road leaves the county and proceeds to Andover. Two good roads run north and south through Salisbury Plain ; one along the Avon valley, while another further west from Devizes, climbing the Plain at West Lavington, passes Tilshead and Shrewton to drop into the Wily valley, and join a third important road from north-west Wilts that has followed the Wily through the west Plain to run to Salisbury.

In the north Wilts low country a great road runs from Highworth south-west to Swindon, and thence west by Wootton Bassett to Malmesbury. In the north-west of the county important roads connect the many towns there, all clustered within a few miles of each other. This is but to mention a few of the principal highways, with which Wiltshire, like most English counties, is so well served, that it is next to impossible for the present generation to imagine what England was like in this respect a little over a hundred years ago.

19. The Roll of Honour of the County.

A selection of the notable men that a county has produced must of necessity be partial and capricious. Many heads, for instance, of the great families from the Tudor period onward, such as the Herberts, the Thynnes, the Longs, and others in Wiltshire, have earned by their well-directed influence, not only in their county but in the country at large, a far stronger claim to be distinguished by posterity than some who are usually included in what it is the fashion to list as the "worthies" of a county. One might go even further back to the mediaeval baronage and episcopate and ask why the Earl of Salisbury and the great countess Ella, who managed the affairs of the Earldom in her husband's absence like a capable man, founded afterwards a great nunnery, and as abbess laid her bones in it, are not more worthy to be "worthies" than Stephen Duck, the "thresher" poet, Poet Laureate, whose crude verses tickled the ear of a not, perhaps, very critical queen (Caroline) and a marchioness (Hertford) with literary affectations, and amused a court for a brief period. The same might be said of Dr Sacheverell, son of a rector of Marlborough, who with scarcely more than average ability and a great deal of impudence made a considerable noise in England for a time as a champion of the Hierarchy and denouncer of the late Revolution in Queen Anne's reign, and was afterwards laughed into

Jane Seymour

obscurity. But such names among county chroniclers find prominent recognition. The poet Thomson, of much worthier fame, is always associated with Wiltshire, as he wrote one of his "Seasons" (Spring) at Marlborough.

Much more intimately connected with the county are some other poets. Sir Philip Sidney, as brother of the talented Mary, wife of Henry Earl of Pembroke, wrote his *Arcadia* at Wilton. Thomas Moore, it is not

Edward Hyde, Earl of Clarendon

generally remembered, lived more than half his adult life in Wiltshire, at Bromham, between Calne and Devizes, in a picturesque cottage still standing, and together with his wife and family lies buried there. The poet Crabbe spent the last period of his life as vicar of Trowbridge and is buried there. The Rev. W. Lisle Bowles, a well-known poet and scholar of the early nineteenth century, lived and wrote as vicar of Bremhill, near Calne, for much of his life. The prose poet of recent times, Richard Jefferies, was essentially a Wiltshireman, son of a yeoman

John Aubrey

farmer at Coate, near Swindon, and drew his inspiration from the soil. Philip Massinger, the dramatist, was born at Salisbury; and Henry Lawes, the well-known musician, who set music to the words of the greatest poets of his day, among others to the *Comus* of Milton, at Dinton.

Richard Hooker was rector of Boscombe, but for little more than a year. Last, but not least, was the saintly poet George Herbert, a cadet of the Pembroke family, whose church and vicarage of Bemerton, near Salisbury, are still standing. He died here in 1633.

Going back to earlier times, Aldhelm, Abbot of Malmesbury and Bishop of Sherborne, was born at the former place, and was famous throughout England for his learning and zeal for education. He is said to have been the first Englishman to write in Latin and to teach Latin verses. He died in the year 709. The famous chronicler William of Malmesbury spent most of his life at that abbey, of which he was librarian. Richard of Devizes is another of the well-known English chroniclers. Edward Hyde, Lord Clarendon, the historian, was born and had his seat at Dinton, near Salisbury. John Aubrey, the delightful antiquary, biographer, and gossip of the seventeenth century, was a Wiltshireman by birth, estate, and residence. Following in his steps, but in far more scientific fashion, came Sir Richard Colt Hoare, the antiquary and historian of the county, which has produced many accomplished workers in this field. John Britton, like Aubrey, born at Kingston St Michael, but in 1771, and of humble parentage, could not be omitted from the most cursory selection of Wiltshire's honoured sons, not so much on account of his 14 volumes on the Cathedrals of England as for his many archaeological works on his native county and his courageous struggle for sixty years devoted to such work against fortune's odds. Thomas Hobbes, the able philosopher and famous defender of the Royal prerogative,

was born near Malmesbury in 1588. Sir Christopher
Wren, the great architect, was the son of a vicar of Knoyle.
Joseph Addison was born at Milston and was member for
Malmesbury, while the witty cleric, Sydney Smith, was

Sir Christopher Wren

curate of Netheravon, on Salisbury Plain, for some time.
Dr Burnet was Bishop of Salisbury for a quarter of a
century, but the great Bishop Poore, who moved the
cathedral from the heights of Old Sarum in the thirteenth
century and rebuilt it on its present site, was essentially a

Wiltshireman. It is interesting to remember, too, that Hazlitt, the celebrated essayist and stylist of the early nineteenth century, used to work for months together in seclusion at the old coaching inn of Winterslow Hut on Salisbury Plain and married from the village of Winterslow. Sir Benjamin Brodie was a native of the same village, where also Henry Fox, afterwards Lord Holland, was born. The Foxes indeed were a Wiltshire family originating in Farley. General Ludlow, the regicide, has been mentioned. Of statesmen Lord Shelburne, the first Lord Lansdowne of Bowood, is perhaps the most conspicuous of modern times, for of living Wiltshire notables we may not speak here. Sir Richard Blackmore, physician to William the Third and poet, was born at Corsham. Henry St John, Lord Bolingbroke, was of the family still seated at Lydiard Tregoze. William Beckford was perhaps one of the best-known men in England at the beginning of the last century, not only on account of his vast wealth, his superb art collections, and the princely house he built at Fonthill, but as the author of the remarkable romance *Vathek*. Archdeacon Coxe, the author of many well-known books of history and travel, was Rector of Bemerton, as was John Norris, a well-known divine in his day. Sir John Davies, the sixteenth century poet, was a native of Tisbury. Thomas Willis, ardent Royalist and afterwards physician to Charles II, was a distinguished anatomist and virtual founder of the Royal Society of London. He was born at Great Bedwyn. General Pitt Rivers of Rushmore as an archaeologist and antiquarian in quite recent times was recognised as an authority of the very first rank. Many of the Pembroke family of Wilton have been eminent, while the Great Protector Somerset, the brother of Henry the Eighth's Jane Seymour, as well as that lady herself, were of course natives of Wiltshire, Wolfhall by Savernake forest being their seat.

20. Administration.

In 1888 and the following years rural government in England, after being eight centuries in the hands more or less of the local aristocracy, whether Norman lords or the later Justices of the Peace, reverted to something more akin to the ancient Saxon system of our ancestors. The general business of a county outside the law courts is now managed by an elected County Council, though the two chief officials of the shire are still the Lord-Lieutenant appointed by the Crown for life, and the High Sheriff who serves for a year. In the Wilts County Council, which sits at Trowbridge, there are sixty councillors and twenty aldermen. This council keeps the main roads and bridges in repair, appoints the police, manages lunatic asylums, and controls many other branches of county government. Besides this, more immediate local control is exercised by Rural District Councils, while the larger parishes have a Parish Council and the smaller a Parish Meeting. Of the former there are 19 in Wiltshire. Some of the towns are boroughs whose business is managed as formerly by mayors and a Town Council. The unchartered towns are under Urban District Councils, of which there are five in Wiltshire. There are nineteen Poor-Law Unions in the county, managed by guardians, whose duty it is to superintend the workhouse and appoint the officers for relieving the poor and aged.

In two large towns, Salisbury and Swindon, the Town

Wanborough Church

Council is the authority for elementary education, but for the rest of the county, including the other boroughs, a committee appointed by the County Council controls it. This committee also controls Secondary Education over the whole of the county, including the two towns above mentioned.

For purposes of justice, besides the petty sessions, held

frequently in all the small towns by borough and county magistrates, there are the quarter sessions, which in Wiltshire are held at Salisbury, Devizes, Marlborough and Warminster. These, too, are composed of county magistrates who elect a chairman from among their number. The most important court for the more serious

The Town Hall, Swindon

cases is that of Assize, presided over by one of His Majesty's Judges. This is held three times a year in Wiltshire, at Devizes and Salisbury. Every case, before being tried at the Assizes, is investigated by a grand jury of magistrates, who decide whether it shall be tried or not. There is also in Wiltshire, as elsewhere, a County Court, presided over by a special judge. This court settles what are known

9—2

as civil cases, that is questions of debt or damages, and can impanel a jury when they involve more than a certain amount of money.

For ecclesiastical purposes Wiltshire is, of course, in the diocese of Salisbury, where the bishop resides, and is divided into archdeaconries, rural deaneries, and parishes. There are five parliamentary divisions in Wiltshire, those of Cricklade, Devizes, Chippenham, Westbury and Wilton, each returning one member to Parliament.

Coate Reservoir, near the banks of which
Richard Jefferies lived

21. THE CHIEF TOWNS AND VILLAGES OF WILTSHIRE.

(The figures in brackets after each name give the population of the parishes in 1901, and those at the end of the sections give the references to the text.)

Aldbourne (1117) gives its name to a chase famous in mediaeval times and much frequented by royal sportsmen. John of Gaunt had a hunting-box here at Upper Upham, still standing, a little fourteenth century manor house elaborated in the Tudor period. Aldbourne church is a very fine one with a grand tower, a beautiful Norman doorway, and a transition Norman arcade. There was a fierce skirmish here in the civil war in which Prince Rupert's horse on their way to the battle of Newbury defeated a Parliamentary force, and extensive remains of the dead were discovered in making a road in the year 1815. (pp. 44, 116.)

Amesbury (1143), a small town on the Salisbury Avon and in the heart of Salisbury Plain; an ancient Abbey foundation, and in the seventeenth century famous for manufacturing the best tobacco pipes in England. It has acquired more importance lately from the neighbourhood of the military camps. It is close to Stonehenge. (pp. 18, 78, 101, 102, 103, 104, 116.)

Avebury (588). The seat of the celebrated prehistoric temple described in the text. The original church of sarsen stone and chalk was said to have been built by a Saxon King for the use of the shepherds. Two of its circular windows replaced in the present church, which is of various dates, are the sole remains of it, with the exception of a curious leaden font of supposed Saxon origin and Norman decoration. Close to the church is an interesting sixteenth century manor house. (pp. 22, 29, 45, 83, 88, 89, 90, 91, 92, 93, 101.)

Great Bedwyn (877). A former pocket borough returning two members to Parliament till the first Reform Bill. Its large cruciform flint church shows Norman and Early English work and

Aldbourne

contains an effigy in full armour of Sir John Seymour of Wolfhall (now a farmhouse in the neighbourhood), the father of Jane Seymour and the Protector Somerset. Chisbury camp is just above the village, on an oval British fortress on the Wansdyke, which passes by here. This is one of the finest camps in the county, the ramparts being still nearly 50 feet high. At Bedwyn, which was important in Saxon times, there was fought in 675 a great battle between Mercia and Wessex, the slaughter being great and the result indecisive. (pp. 75, 94, 101, 116, 128.)

Box (2405), on the eastern edge of the county and intimately concerned with the adjacent quarrying industry. The longest railway tunnel in England except that of the Severn runs under its hill. (pp. 25, 42, 72.)

Bradford-on-Avon (4514), an old centre of the cloth trade, now practically extinct here, and picturesquely situated on the slope of a hill above the Bristol Avon. It is noted for its numerous very old houses, and also for the most perfect little Saxon church in England. (pp. 11, 12, 25, 28, 42, 58, 71, 96, 97, 100, 103, 109, 110, 114, 115.)

Brinkworth Church

Broad Chalk (629). Has some importance as the chief village in the long secluded Vale of Chalk and as the residence for many years of John Aubrey the celebrated Wiltshire antiquary. There is a large cruciform church, part Early English and part Perpendicular, in which is a remarkable memorial window to the famous Welsh scholar and theologian, Rowland Williams, who was once Vicar here.

Bromham (1136) lies between Melksham and Devizes and is the burial place of Thomas Moore and several members of his family. The poet lived at Sloperton Cottage in the parish for the last 35 years of his life. The church is large, mainly Perpendicular in character, and contains a rich and handsome chapel of Henry VII age, associated with the Bayntons, an ancient and powerful Wiltshire family who entertained kings in their great mansion here, which was destroyed in the Civil War. (pp. 13, 124.)

Bulford (1386). Though there is an old church and small village here its importance is derived from the permanent camp on the downs in the neighbourhood. (p. 68.)

Calne (3457), a borough market town situated beneath the north-west corner of the Marlborough Downs and chiefly noted for its bacon factories. A great council was held here in 978 to decide the issue between the celibate and the married clergy, Archbishop Dunstan supporting the former and Bishop Beornhelm of Winchester the latter. The floor of the council chamber collapsed with the entire assemblage excepting the Archbishop, who is said to have caught on a beam. Calne was a pocket borough of the Lansdowne family at Bowood in former days, but distinguished for the eminent men who represented it, Lords Macaulay and Sherbrooke among them. Bowood Park adjoins the town. (pp. 17, 22, 24, 25, 26, 28, 29, 41, 58, 68, 71, 91, 94, 95, 124, 125.)

Castle Combe (357) is in the Cotswold country and has the reputation of being the most naturally beautiful village in Wiltshire. It has also been the home of the Scropes from the time of the Lord Chancellor of Richard II. It is surrounded by wooded hills on one of which are the remains of a castle of the Dunstervilles. There is a fine market cross and many very beautiful old houses of Cotswold stone and style. The site of the Norman castle is an ancient British camp, and in the immediate neighbourhood a Roman villa was unearthed in the last century. The beautiful tower alone remains of a fifteenth century church, built at a time when the clothiers of this part of the country were growing rich. (pp. 25, 36, 116.)

Cricklade Church

Chilseldon (1136) stands picturesquely on the northern slope of the Marlborough Downs above Swindon. The church is not especially interesting though full of plain tablets and memorials, many of them to the old family of Calley who have been the principal landowners for some centuries.

Chippenham (5074) is a borough market town on the Bristol Avon and a railway junction on the Great Western Railway main line. It has a large market in which "double Gloucester" cheese is an important item. In former days it was one of the chief cloth-weaving towns. It is a very ancient place, having been a market town and a frequent residence of the Kings of Wessex in Saxon times. Its name is derived like that of "Cheapside" from the Anglo-Saxon *ceapan*, to buy. It was a chief seat of King Alfred's Danish wars of the ninth century. Maud Heath's Causeway leads from Bremhill to Chippenham. (pp. 25, 26, 28, 41, 58, 62. 66, 71, 76, 120, 132.)

Corsham (2600) is an ancient little town, and was the occasional residence of Wessex Kings. There are limestone quarries in the neighbourhood. Corsham Court (Lord Methuen) is a fine specimen of the Elizabethan style. (pp. 42, 71, 72, 111, 112, 128.)

Cricklade (1500) is the only Wiltshire town upon the Thames, or as here called the Isis, which is navigable for boats to this point. It is the centre of an agricultural district and a very ancient place, being situated on the Roman road from Newbury to Cirencester. It has a fine cruciform church. An absurd tradition, encouraged by the poet Drayton, associated its name with a colony of Greek philosophers who settled here before the Roman invasion, whereas it has a Celtic derivation. (pp. 12, 25, 101, 103, 120, 121, 132.)

Devizes (6532). In former days and still occasionally called "The Vies," its ancient name being "ad devisas" indicating its position on some boundary line. It stands 500 feet above the sea on the western edge of the Marlborough Downs and in the heart of the county. The Assizes are held here as well as at Salisbury. The county lunatic asylum and the depot of the Wiltshire regiment are also here. It has always from its situation been an important market town. In mediaeval times it was famous for its large and powerful castle and is now distinguished for the architecture of its two noble old churches. Brewing, malting, and the manufacture of snuff are active modern industries. The father of Sir Thomas Lawrence the great painter kept the Bear Inn here, which is still one of the chief hostelries on the Bath road.   (pp. 9, 17, 24, 29, 35, 36, 59, 78, 81, 82, 88, 100, 103, 105, 120, 121, 124, 126, 131, 132.)

Donhead St Mary (1121).   This village, adjoining Donhead St Andrew, has a church of some distinction, containing Early English, Decorated, and Perpendicular work, and is most picturesquely situated beneath high hills.   Close by, in Lord Arundel's park, are the ruins of Wardour Castle, which stood a siege in the Civil War, under the command of its châtelaine.

Downton

Downton (1786), a large village by the Avon on the southern edge of the county which formerly returned two members to Parliament.   It contains several buildings of historic association, including a supposed Saxon Moot Hill and a house once connected with Sir Walter Raleigh's family.   Here too is one of the finer churches of the county, cruciform, with central tower and ranging in style from Norman to Perpendicular. (pp. 25, 30, 96, 101, 115.)

Edington (779), near Westbury, is celebrated for its unusually large and beautiful church, built by a Bishop of Winchester, a native of this village in 1352, and considered one of the finest specimens of transition from Decorated to Perpendicular in England. It is cruciform with a central tower, and contains many interesting monuments. A college and Priory were founded here concurrently with the church of which scant traces yet remain in an adjoining farmhouse. In Jack Cade's rebellion in 1449, Ayscough, Bishop of Salisbury, was dragged from the altar by peasants and murdered in an adjacent field. The camp on the summit and the great white horse carved on the slope of Bratton Down are held to mark the site of Alfred's great victory over the Danes mentioned in the Chronicle as Ethandun. (pp. 20, 76, 103.)

Heytesbury (699). A large village near the head of the Wily valley. It has an interesting collegiate cruciform church of Early English mingled with later Perpendicular style, and a fifteenth century hospital or almshouse founded by the great Wiltshire family of Hungerford, who owned this property before going to Farleigh. Close by on the downs above are the noted camps of Scratchbury, Battlesbury, and Knook. (p. 116.)

Highworth (2046), an old town on the north-east border of the county. It stands on a high ridge and is purely an agricultural centre, but of small importance. There is a fine Perpendicular church around which a skirmish was fought in 1645 and a second battle a few weeks later. (pp. 12, 24, 25, 46, 116, 121.)

East Knoyle (814) is the birth-place of Christopher Wren, whose father was Rector here and also Dean of Windsor. (p. 127.)

Lacock (1159). A large village containing an unusual number of fifteenth and sixteenth century houses, but chiefly famous for the adjoining Abbey on the banks of the Bristol Avon, founded as an Augustine nunnery by Ella Countess of Salisbury in 1232 who became its Abbess. Large portions of the nunnery in Early English and Perpendicular style remain in the present house and outbuildings, and are among the most complete remains of a monastic building in the county. The Abbey and the village taken together are perhaps the most suggestive picture of a former day that exists in Wiltshire. (pp. 78, 102, 103, 104, 116.)

Ludgershall (576) was formerly only a country village on the south-east corner of Salisbury Plain but as a station and entrepôt for the military camps is now growing in importance. There are also here the fragments of a castle which often went in the matter of government with the neighbouring Royal castle of Marlborough. (pp. 105, 106, 121.)

Malmesbury (3133). That portion of the Abbey now used as a parish church—six bays of the original nine of the nave—is sufficient testimony to the former grandeur and scale of the great cruciform building which surpassed that of many cathedrals. It was erected early in the twelfth century by that powerful prelate and active builder, Roger of Salisbury, in the transition Norman style. With its massive circular pillars and lofty clerestory and triforium the interior is still very majestic. Originally, in Professor Freeman's opinion, it was no doubt one of the grandest in England. The profusely sculptured and heavily moulded Norman south porch is one of the remaining glories of the church. Two great towers fell at different periods of the sixteenth century, doing infinite damage. The site of the church on a rocky eleva-tion between the Marden and the Avon adds to its dignity. William of Malmesbury, the celebrated chronicler, was librarian of the monastery, of which no traces remain *in situ*. King Athelstan was its munificent patron, who also left to the town a considerable tract of land, which to this day is enjoyed as free allotments by the townspeople under the original and favourable

Market Cross, Malmesbury

conditions. There is a fine specimen of a market cross in the Perpendicular style. (pp. 12, 25, 28, 78, 80, 100, 103, 104, 116, 121, 126, 127.)

Market Lavington (978). A picturesque little town of a single street, once a place of cloth manufacture, now of purely agricultural connection and situated beneath the northern rampart of Salisbury Plain.

Malmesbury Abbey, West Front

West Lavington (1027). A mile from the above, also a picturesque old-fashioned village at the foot of the down. In the church is a monument to Captain Penruddock, who was murdered by Cromwellian troopers in a house still standing here. This village is at the northern entrance of a main road through Salisbury Plain. (p. 121.)

Marlborough (3300), a borough town of great coaching importance in former days and formerly a Parliamentary borough.

It stands on the Kennet and on the Bath road, and is now chiefly associated in the public mind with the well-known school founded here in 1843. This occupies the site of the royal castle of mediaeval times and the actual mansion of the later Seymours who succeeded to the property. Though a local market town it owes much of its present prosperity to the college. Marlborough is remarkable for its wide sloping High Street and picturesque

Silbury Hill

appearance and situation, though much of the town, including the greater part of its best church, St Mary's, was destroyed by a terrible fire in the Commonwealth period. It is now served by two lines of railroad and is adjacent to Savernake forest, Avebury, and Silbury. There is also here an ancient and well-known grammar school, now reconstructed upon different lines. The site of Cunetio, an old Roman town, lies a mile to the eastward and Savernake forest adjoins the town. (pp. 5, 21, 22, 26, 29,

39, 55, 78, 80, 88, 89, 91, 93, 95, 98, 103, 104, 105, 111, 115, 117, 120, 121, 123, 131.)

Melksham (2450) stands on the Bristol Avon, it possesses some woollen, horsehair cloth, sacking, and rope industries, and also some engineering works. It has lately increased in prosperity and population. (pp. 25, 41, 58, 66, 71, 120.)

Mere (2000) is on the south-western edge of the county and forms part of the Duchy of Cornwall. It still produces a little coarse linen and silk. (p. 103.)

Pewsey (1722). A small town on the head waters of the Salisbury Avon, which gives its name to the vale intervening between Salisbury Plain and the Marlborough Downs, and is the centre of this considerable agricultural district. It has some reputation for the creditable energy which makes the ancient institution of "Pewsey Feast" an annual gathering of much note. The church is large and interesting with a Perpendicular tower and a good thirteenth century nave. The fine escarpment of Martinsell with a British village and well-defined camp on its slope and crown respectively, with Huish hill, just to the west, also carrying a British village, overlook Pewsey from the north. On the Salisbury Plain side are Easton hill with a circular camp and British village, also Milton and Pewsey hills, where are more traces of British villages. (pp. 18, 20, 47, 120, 121.)

Potterne (1150), a large and picturesquely seated village near Devizes containing some very good specimens of half-timbered houses and a very beautiful early English cruciform church with a square central tower. The church with its extreme severity, beautiful proportions, and triple lancet windows has something suggestive of Salisbury Cathedral and is thought to have been possibly built by the same founder, Bishop Poore. (pp. 39, 101, 110.)

Preshute (1622). An outlying parish of Marlborough containing several hamlets along the Kennet and outlying houses connected with the College. The church still preserves its Norman pillars, though partially rebuilt, and contains a noted font of black basalt in which tradition says King John was baptised, the truth probably being that two or three of his children were, since that monarch was much at Marlborough Castle, from the chapel of which the font came. One of the several white horses of Wiltshire is cut in the chalk of the down in this parish, and here also is the principal cromlech in the county, popularly known as the "Devil's Den." Near by are the valleys or denes most noted for the accumulation of the sarsen stones or "Grey wethers." (p. 100.)

Purton (2525). A large village in the vale of Swindon. The church is cruciform and Perpendicular in character but chiefly distinguished for its two steeples. One is in the centre and carries a stone spire, the other is at the west end and has open battlements. Lord Chancellor Hyde's family came from Purton though he himself was born at Dinton near Salisbury. Their house is still standing. According to Aubrey, Anne Hyde, mother of the Queens Mary and Anne, was born here. The Manor belonged to the Maskelyne family and Dr Maskelyne, Astronomer Royal in the eighteenth century, is buried here. Purton fair was once famous. (p. 104.)

Ramsbury (1779). A large agricultural village on the Kennet, the seat of the Bishops of Wiltshire from early in the tenth to the middle of the eleventh century when, after a brief transfer to Sherborne, the united diocese of Wilts and Dorset was centred at Salisbury. During the negotiations between William of Orange and the King's commissioners on the Prince's march to London, the commissioners took up their quarters at Ramsbury Manor in the predecessor of the present house, while William lay at Littlecote two miles down the Kennet. At the latter place an agreement which greatly influenced the future of England was arrived at. Littlecote is one of the finest surviving Tudor houses in England and is associated with a dark deed in the sixteenth century, of sufficient note to be mentioned in Macaulay's *History of England*. One of the finest Roman pavements ever discovered in England was found here, but was unhappily destroyed. (pp. 29, 38, 104, 112.)

Salisbury (17,000). The capital of Wiltshire and cathedral city of the diocese which includes Wilts and Dorset. Its cathedral is the finest specimen of pure Early English architecture and one of the largest in England. Its spire is quite the highest, being over four hundred feet. The building was commenced by Bishop Poore in 1220 and completed about 1260; the spire being added in the next century. The shape of the cathedral is a double cruciform, the western transepts however being considerably the longest. Its extreme length is 473 feet and the measurement round the exterior of its walls just half a mile. The fabric generally has not been materially altered by either injuries or restoration and, having been erected all at one time and upon definite plans, presents both inside and out a splendid and complete picture of beautifully proportioned Early English work. There are a great many interesting monuments, but the interior details of the building have much suffered from unwise restoration. The cloisters, which are of later date, are among the finest in England, while the chapter house is a striking octagonal building of the time of Edward I, and noted for its wealth of internal sculpture illustrating mediaeval conceptions of religious subjects and scenes from the Old Testament. Splendid as the fabric is, a certain excess of severity and lack of colouring detract something from its interior, though the surrounding close and precincts with their many beautiful old houses have no superior in any cathedral town. The Deanery, the King's House, and the King's Wardrobe are the best in the close. There are a few manufactures in the place, mainly connected with agriculture, of which Salisbury is an important centre and market. Its streets contain many interesting old buildings, and it lies picturesquely on the levels where the valleys of the Avon and the Wily just united with the Nadder meet. The former stream flows through the heart of the town. This together with the cathedral was moved down to the present site from the adjacent heights of Old Sarum (or to speak more accurately rebuilt) in the thirteenth century. Its streets were then laid out in right angles, a form they preserve to this day in a manner unusual in ancient towns. The site of the former town and capital of Old Sarum is now but a desolate, intrenched and lofty hill-plateau overlooking its successor. (pp. 3, 12, 17, 18, 19, 29, 30, 31, 37, 46, 47, 65, 73, 78, 79, 82, 88, 89, 100, 101, 105, 110, 111, 115, 117, 120, 121, 125, 126, 127, 130, 131, 132.)

Hinton Parva Church, near Swindon

Great Sherstene (1359) is supposed to be the scene of a great battle between Edmund Ironside and the Danes under Canute. Much of the village and a large Norman church are built within an old intrenched camp on a promontory between two streams. There is another camp close by, both being probably Saxon and connected with the same events.

Steeple Ashton, literally **Staple Ashton** (650). This was once a settlement of cloth manufacturers, which was burnt in the sixteenth century and its trade transferred to Market Lavington. A ball upon the top of a column erected over two centuries ago marks the site of the old market cross in the village street. The beautiful large church (late Perpendicular) was erected at the end of the fifteenth century in part by two wealthy clothiers, Walter Lucas and Robert Long, the latter of the well-known Wiltshire family whose present seat of Rood Ashton is near by.

Standlynch on the Avon

Swindon (45,000) now consists of Old and New Swindon and chiefly depends on the Great Western Railway works which are situated there. It is much the largest town in Wiltshire and is a borough. It is also the only town in the county which has made great strides in population. Within easy memory it con- sisted of Old Swindon only, an ancient borough and market town with a population of about three thousand. It lies beneath the northern slope of the Marlborough Downs at the western end of the Vale of the White Horse. The population by birth and descent is largely derived from outside the limits of the county. It is an important market town, particularly for cattle. (pp. 13, 15, 17, 22, 24, 25, 36, 41, 53, 57, 66, 71, 72, 75, 120, 121, 125, 130, 131.)

Tisbury (1464), a large village and capital of the vale of Wardour stands above the river Nadder. The church is one of the largest in the county, most of it being sixteenth century work. Some windows of later date still are assigned to Sir Christopher Wren. Sir John Davies, the sixteenth century poet, was born here. Close by is Wardour Castle, the seat of Lord Arundel, and many monuments belonging to this ancient family are in the church. Near by is Fonthill Abbey, so renowned in the eighteenth and early nineteenth centuries as the magnificent home of Beckford, the author of *Vathek*, one of the wealthiest men of his day. The Tisbury quarries of building-stone are well known. (pp. 41, 53, 80, 128.)

Trowbridge (11,526). Stands on the Biss, a small tributary of the Bristol Avon. It was in former days one of the principal seats of the cloth trade, a little of which still survives. It is now an important railroad and market town and the headquarters of the Wilts County Council. The old church, St James's, like that of Chippenham, has been almost re-built in restoration. The Rev. George Crabbe the poet was Vicar here from 1814 to 1832, and lies buried in the chancel. The Rectory which he inhabited is a fifteenth century building of great interest. (pp. 12, 13, 23, 25, 28, 58, 66, 71, 78, 103, 120, 124, 129.)

Wanborough (806). The church, very finely situated with tower, nave, and chancel of Perpendicular character, is the second of the two instances in Wiltshire of a double spire, the lesser one in this case being on the east end of the nave. Wanborough

(Wodens burh) was the scene of two fierce battles in the Saxon period, the first in 501 when Ceawlin was defeated by his nephew Cerdic, the second in 714, an indecisive one between Ina of Wessex and Ceolred of Mercia. Many Roman roads crossed at or near Wanborough, and strategically it was one of the keys of Wessex. On the down above is Liddington camp, enclosing eight acres within a well-preserved rampart, still some forty feet high from the bottom of the ditch. This was a favourite haunt of the naturalist-philosopher Richard Jefferies, whose home and that of his fathers before him at Coate is easily visible in the vale below. (pp. 75, 104, 130.)

Warminster (5547). An ancient town which derives its name from the minster which once stood here on the banks of the Were. Its present buildings however are mostly modern and uninteresting though the situation in a valley between the bold escarpments of Salisbury Plain with their intrenched crests and Cley hill is very striking. In the old grammar school here the famous Dr Arnold of Rugby received his earlier education. This is mainly an agricultural centre and market town with a malting and corn trade. The curious isolated Cley hill (900 ft.) rises above Warminster where was a beacon in the time of the Spanish Armada. Longleat, Lord Bath's magnificent seat and park, is also near. The house, which is a mixture of the English and Italian style, was erected in the sixteenth century and was then one of the finest in England. (pp. 12, 17, 19, 25, 29, 35, 36, 53, 131.)

Westbury (3300), like Warminster, stands beneath the western escarpments of Salisbury Plain and is an old market town of small importance to-day. Like most of its neighbours it was once busy with the cloth-trade. A vein of iron ore close to the town is now worked. Westbury once returned two members to Parliament. The railway down the Wily Valley to Salisbury branches from here. The church is cruciform in shape and

originally of Norman style, but like a majority of Wiltshire
churches, was much altered in the Perpendicular period, which
coincided with the prosperity of the mediaeval wool-trade. There
is a conspicuous monument in it to Sir James Ley, who was made
Earl of Marlborough by Charles I when he became President of
the Council.   There is also a chapel of Henry VI period of the
Willoughby de Broke family. Above on Bratton Down is the
only ancient, though re-cut, "white horse" in Wiltshire. (pp. 12,
13, 17, 18, 23, 25, 35, 36, 47, 58, 72, 76, 120, 132.)

Wroughton Church

Wilton (2200), the ancient capital of the county and of
Wessex. It stands at the junction of the Nadder and the Wily
and for long was a town of great importance. It is only three
miles from Salisbury and adjoins the celebrated seat of the
Earl of Pembroke, Wilton House. Here is the well-known
carpet-factory, the first establishment of this industry in England.

In 823 the forces of Wessex under Egbert overthrew those of Mercia in a decisive struggle for supremacy, and fifty years later Alfred defeated the Danes here, who, after half a century, had their revenge in burning the place down. Wilton was a very important town in all matters till the year 1244, when the road leading from Salisbury to the west was carried another way by a bridge over the Avon, which injured it fatally. The old church is a ruin, the present one is a gorgeous Lombardic building erected by the late Lord Herbert of Lea. Bemerton, with the church and vicarage occupied by George Herbert, is about a mile down the river towards Salisbury. (pp. 2, 19, 29, 55, 71, 73, 75, 78, 103, 104, 111, 124, 129, 132.)

Winterslow (800), on the edge of Hampshire, is the former home of the Fox family. Hazlitt the essayist spent a part of several years at Winterslow Hut, a famous old coaching inn, and married in the village. Sir Benjamin Brodie was born here, while his father was Vicar. A successful cooperative experiment in small proprietorship of houses and land has been going forward here for over a dozen years. (p. 128.)

Wootton Bassett (2258). A small market town of one long street occupying the point of a ridge at the western end of the vale of Swindon, with a half-timbered town-hall and a market house on stone pillars in the middle of the street. It is the centre of a milk and cheese country and was once a Parliamentary borough. (pp. 24, 25, 36, 41, 70, 120, 121.)

Wroughton (2225). A village three miles south-west of Swindon, with a church dating back to the eleventh century.

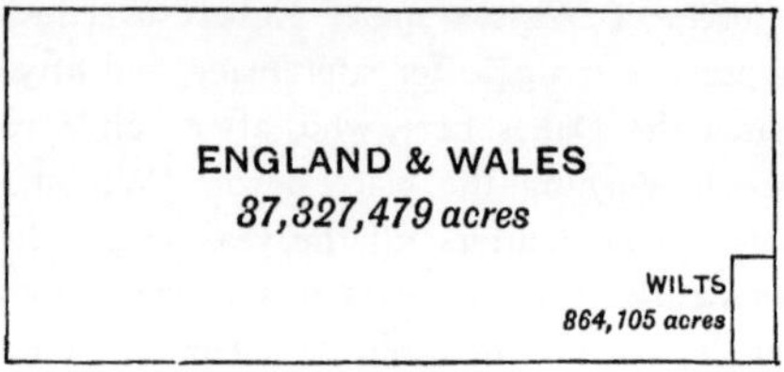

ENGLAND & WALES
87,327,479 acres

WILTS
864,105 acres

Fig. 1. The Area of Wilts compared with that of England
and Wales

ENGLAND & WALES
32,527,843

WILTS
271,394

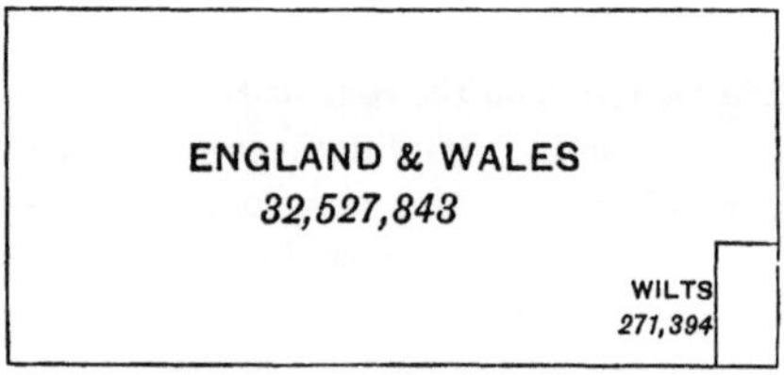

Fig. 2. The Population of Wilts compared with that of
England and Wales

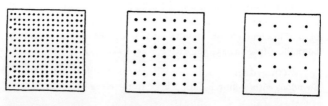

Lancashire 2347   England and Wales 558   Wiltshire 201

Fig. 3. Comparative Density of Population to the
Square Mile (1901)

(*Note. Each dot represents* 10 *persons*)

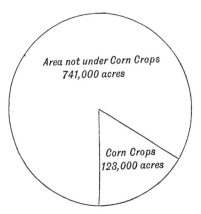

Fig. 4. Proportionate area of Wilts under Corn Crops

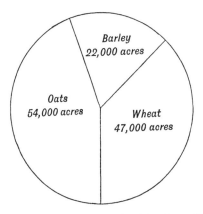

Fig. 5. Proportionate area of cultivation of Wheat,
Oats, and Barley in Wilts

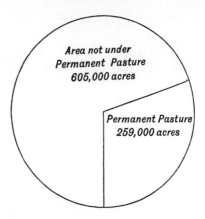

Fig. 6. Proportionate areas under Permanent Pasture and
Not under Permanent Pasture in Wilts

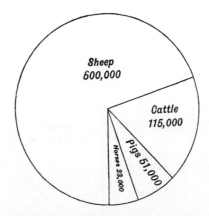

Fig. 7. Proportionate numbers of Live Stock in Wilts

www.ingramcontent.com/pod-product-compliance
Ingram Content Group UK Ltd.
Pitfield, Milton Keynes, MK11 3LW, UK
UKHW042144280225
455719UK00001B/101